T0235809

ELECTROLYTES AT INTERFACES

Progress in Theoretical Chemistry and Physics

VOLUME 1

Electrolytes
at Interfaces

by

S. Durand-Vidal

J.-P. Simonin

and

P. Turq

Université Pierre et Marie Curie,
Laboratoire d'Electrochimie,
Paris, France

KLUWER ACADEMIC PUBLISHERS
DORDRECHT / BOSTON / LONDON

Library of Congress Cataloging-in-Publication Data

ISBN 0-7923-5922-4

Published by Kluwer Academic Publishers,
P.O. Box 17, 3300 AA Dordrecht, The Netherlands.

Sold and distributed in North, Central and South America
by Kluwer Academic Publishers,
101 Philip Drive, Norwell, MA 02061, U.S.A.

In all other countries, sold and distributed
by Kluwer Academic Publishers,
P.O. Box 322, 3300 AH Dordrecht, The Netherlands.

Printed on acid-free paper

Contents

III   Specific applications   125

Foreword

The behavior of ions in solutions has been progressively understood by the efforts of great pioneers: Debye, Onsager. Soon after, Verwey, Overbeek, Landau, Derjagin, established a basic picture of elastic properties near an interface. Later came the discussion of transport phenomena, normal or parallel to the interface: the present authors have contributed significantly to this field. They combine (at least) three forms of culture: (i) a practical knowledge of the relevant physical chemistry: colloids, polyelectrolytes, etc.; (ii) a broad view of the experimental techniques required to probe the statics and the dynamics; (iii) a strong practice of computational methods derived from the theory of liquids: calculation of correlation functions, incorporation of backflows, simulations. These calculations require a lot of care, but they do provide the tools required to describe many subtle effects: the deformation of the ionic cloud around a moving object is one fascinating example.

This book is not easy: it requires an attentive reader. But such a reader, if he is interested in a novel feature, will find in the text all the theoretical instruments which he needs. And there is a lot to think about: on nanoparticles, on polyelectrolytes, on mixed systems with polymers and colloids, we are still learning surprising facts every year. This book will help some of us to understand them better.

P.G. de Gennes

Paris, December 1998

Aim and Scope

We intend to provide a modelistic and modern discussion of electrolytes at interfaces aimed at theoretical physical chemists, chemists and theoretically oriented experimentalists in these areas - we will discuss primarily the structural and thermodynamic properties, in relation to simple and also realistic model.

We will also summarize the status of the experimental knowledge of the chemical physics of these systems. The current and open problems of the field, will be analysed in detail taking into account the recently published experimental and theoretical work.

There has been, in recent times, a big surge in the amount of experimental and theoretical works in the areas of interfacial electrochemistry, colloidal chemistry, polyelectrolytes and Langmuir-Blodgett films.

The understanding of the structural aspects of this rather diverse body of data relies on simple models, which work well in some cases, and are still insufficient for others.

The basic idea of embedding a hard core into the traditional theories like the Debye-Hückel theory, (bulk electrolytes or Gouy-Chapman theory, metal-electrolyte interfaces), leads to tractable, and in some cases (colloidal suspensions) good theories, provided that we can ignore the effect of the structure of the solvent by using a continuum dielectric model (primitive model).

When the molecular structure of the solvent plays an important role, then the situation is not so clear, and there is a genuine need for further theoretical development.

An elementary way of taking into account the granularity of the solvent is the hard sphere-point dipole model, which accounts for the excluded volume effects and part of the polarization effects of the solvent. Unfortunately, even with modern vectorized computers, it is difficult to simulate these systems, even in the bulk phase.

We discuss the exact sum rules for the more complex, and realistic models, such as the contact theorem, which gives the amount of ions in the interface as a function of the excess charge, and the screening theorems, which are conditions on the charge distributions in an inhomogeneous system.

We intend to discuss also dynamic aspects related to transport processes such as diffusion and electrophoresis, however, at a more phenomenological level.

We will also discuss some important new experimental aspects such as the Israelachvili experiments measuring the forces between two plates, as well as other experiments in which the forces between colloidal particles are directly measured by direct optical microscopy.

We are deeply indebted to Professor Lesser Blum who was closely involved in a first version of this book.

We are grateful to O. Bernard for the communication of several personal documents and for his helpful discussions and to L.-H. Jolly for his devotion to the practical and technical achievement of this book.

The Authors

Part I

Basic Concepts

Chapter 1

Hydrodynamic properties

The description of the transport properties of electrolyte solutions requires some basic information on the hydrodynamic interactions between the solute particles.

This chapter is aimed at giving a concise presentation of the necessary tools. The first section of this chapter is devoted to the basic principles of hydrodynamics. In the second section, a description of hydrodynamic interactions between moving particles in a fluid is presented. Limitation is made to the level of the Navier-Stokes theory commonly used in the theory of electrolyte solutions.

1.1   Introduction

Two questions may be asked at the beginning of this chapter:

- What is a hydrodynamic interaction ?

- To what extent must it be taken into account ?

The first question can be answered by the presentation of a simple experiment. Consider a rigid spherical particle of macroscopic size (radius R) and mass m, immersed in a large volume of a fluid. The fluid and the sphere are at rest. The fluid viscosity is denoted by η.

At time $t = 0$, a force $\mathbf{F}_{app}$ [1] is applied to the sphere. The latter is therefore accelerated, according to Newton's law, until it reaches a constant velocity. Then the applied force is balanced by a friction force $\mathbf{F}$

$$\mathbf{F} + \mathbf{F}_{appl} = 0$$

The Stokes's law states that the force $\mathbf{F}$ is proportional to the velocity $\mathbf{v}$ of the sphere and to its size. It reads

$$\mathbf{F} = -6\,\pi\,R\,\eta\,\mathbf{v} \tag{1.1}$$

If another sphere moves in the opposite direction on a rectilinear trajectory parallel to the first sphere it will interact with the first sphere and cause a decrease in the velocity $\mathbf{v}$ if the two spheres are close enough, at a distance smaller than a few radii R.

This interaction is what may be called a "wake effect". It is well known from common experience, as in the case of a fast truck passing a cyclist or a boat coming in the vicinity of a launch.

Based on this observation a preliminary answer can be given to the second question: wake effects should be taken into account when the particle concentration corresponds to a mean interparticle distance D of only a few particle diameters d. The critical concentrations C_{cr} and distances D_{cr} may therefore be related by

$$C_{cr} = D_{cr}^{-3} \tag{1.2}$$

$$D_{cr} = nd \tag{1.3}$$

where n is a small number depending on the solvent viscosity η, which is an enhancing factor for the wake effect (see the Section "Microscopic origin of viscosity" below). Moreover, the number n should be independent of the mean particle velocity if relative velocity changes are considered.

A rough estimation for aqueous solutions shows that $n = 3$ is a reasonable assumption, so that we get

$$\mathcal{O}(C_{cr}) = (3d)^{-3} \tag{1.4}$$

with the symbol $\mathcal{O}(X)$ meaning an estimation of the order of magnitude of a quantity X. Taking $d = 8$ Å for the hydrated ion diameter yields

$$\mathcal{O}(C_{cr}) = 0.1 \text{ mol dm}^{-3}$$

The distance D_{cr}, here $D_{cr} \simeq 25$ Å, seems a sound estimation for the cut-off distance above which wake effects may be neglected in water.

---

[1] vectors will be denoted by bold type characters throughout

1.2 General aspects of hydrodynamics

In hydrodynamics the description of fluid properties is based usually on a continuum hypothesis. Comprehensive treatments are given in Refs. [1-3] The elementary volumes have side lengths in the micrometer scale and hence are very large in comparison with the particle sizes. Such elementary volumes contain about 10^{10} particles forming a continuum.

Two methods are commonly used to study a fluid, the Lagrangian and the Eulerian picture. The Lagrangian description considers the trajectories $\mathbf{r}(t)$ of "fluid points" as a function of time, which would be observed if very small iron filings particles were present in the fluid. Although intuitive, this description leads generally to a rather cumbersome analysis and is not the more appropriate one. In lieu of this approach the Eulerian method is preferred, investigating the motion of the fluid at fixed points of the space as a function of time. In this theory the velocity field $\mathbf{v}(\mathbf{r}, t)$ is the basic studied quantity.

Here, only incompressible fluids will be considered. This model is sufficiently close to real liquids and solutions.

1.3 Inviscid fluids

The inviscid fluid, though an idealized system, yields a suitable reference frame in which real fluids can easily be taken up. This model is in use since the 18th century. It has achieved the first step for the extension of classical mechanics to deformable media.

In the particular case of incompressible fluids the continuity equation, given by the relation

$$\frac{\partial \rho}{\partial t} + \text{div}(\rho \mathbf{v}) = 0 \qquad (1.5)$$

where div denotes the divergence operator, is greatly simplified since the density ρ is a constant. One gets

$$\text{div}\,\mathbf{v} = 0 \qquad (1.6)$$

This equation means that, as the magnetic field in electrodynamics, the velocity vector has no sources in incompressible fluids.

1.3.1   Time derivative of velocity

Following the motion of a fluid, the increment of velocity $d\mathbf{v}$ in the Eulerian specification results not only from a local increase at point $\mathbf{r}$ but also from the displacement of the elementary volume of the fluid during the interval of time dt. Taking into account the defining relation of velocity

$$\mathbf{v}\, dt = d\mathbf{r} \tag{1.7}$$

yields the relation

$$\frac{d\mathbf{v}}{dt} = \frac{\mathbf{v}(\mathbf{r} + \mathbf{v}dt, t + dt) - \mathbf{v}(\mathbf{r}, t)}{dt} = \frac{\partial \mathbf{v}}{\partial t} + (\mathbf{v}.\nabla)\,\mathbf{v} \tag{1.8}$$

where the operator $(\mathbf{v}.\nabla)$ is the scalar product of the velocity vector by the gradient operator. In cartesian coordinates (x, y, z) this product is given by the expression

$$\mathbf{v}.\nabla = v_x \partial_x + v_y \partial_y + v_z \partial_z$$

with the notation

$$\partial_x \equiv \frac{\partial}{\partial x}$$

Similarly to eq 1.8 the following relation can be obtained

$$\frac{d}{dt} = \frac{\partial}{\partial t} + \mathbf{v}.\nabla \tag{1.9}$$

1.3.2   The Euler's equation

The net force on a volume V of an inviscid fluid, bounded by a surface S, arising from a pressure gradient is

$$\mathbf{F}_V \equiv -\iint_S p\,\mathbf{n}\ dS \tag{1.10}$$

with p the pressure on the surface and $\mathbf{n}$ the unit vector perpendicular to the surface locally and oriented outwards. This equation can be transformed by use of the Ostrogradski's theorem to yield

$$\mathbf{F}_V = -\iiint_V \nabla p\ dV \tag{1.11}$$

This equation means that the force per unit volume of fluid is simply $-\nabla p$.

Applying the Newton's law gives the Euler's equation: consider an elementary volume of fluid of density ρ experiencing a pressure gradient and placed in an external field of intensity $\mathbf{g}$ per unit of mass; then we have that

$$\rho\frac{d\mathbf{v}}{dt} = -\nabla p + \rho\mathbf{g} \tag{1.12}$$

Eq 1.12 is valid for an ideal inviscid liquid. It was established by Euler in 1755. Using eqs 1.8 and 1.12 leads to

$$\frac{\partial\mathbf{v}}{\partial t} + (\mathbf{v}.\nabla)\,\mathbf{v} = -\frac{1}{\rho}\nabla p + \mathbf{g} \tag{1.13}$$

This equation constituted the first main equation of hydrodynamics.

1.3.3 Bernoulli's theorem

An interesting class of flows is that of stationary flows, for which

$$\frac{\partial}{\partial t} \equiv 0$$

Using the vectorial identity

$$(\mathbf{v}.\nabla)\,\mathbf{v} = \nabla(v^2/2) - \mathbf{v} \times \mathbf{curl}\,\mathbf{v} \tag{1.14}$$

where the symbol $\times$ denotes the vector (also: outer or cross) product.

Then Euler's equation reduces to

$$\mathbf{v} \times \mathbf{curl}\,\mathbf{v} = \nabla(\frac{\mathbf{v}^2}{2} + \frac{p}{\rho} + U) \tag{1.15}$$

when taking into account that ρ is constant for incompressible liquids and $\mathbf{g}$ can be represented as the gradient of a potential U, $\mathbf{g} = -\nabla U$.

Following a streamline, *i.e.* a line such that the velocity is tangent to this line at each point, eq 1.15 can be multiplied by the unit vector, that is tangent to the streamline, to yield

$$\frac{\partial}{\partial s}(\frac{\mathbf{v}^2}{2} + \frac{p}{\rho} + U) = 0$$

where s is the curvilinear abscissa along the line, from which Bernoulli's theorem is deduced: the quantity

$$H = \frac{v^2}{2} + \frac{p}{\rho} + U \tag{1.16}$$

is a constant along each streamline.

From eq 1.15 also follows that the quantity H is constant all throughout the fluid if **curl v** is zero everywhere in the liquid. In particular, if the action of gravity can be neglected, if for instance the pressure gradient is dominant, then the quantity $(v^2/2 + p/\rho)$ is a constant showing that the points where the pressure is maximum are the stagnation points where $\mathbf{v} = \mathbf{0}$.

1.3.4   Vorticity

Upon application of the **curl** operator to eq 1.13, and with the use of eq 1.14, one gets

$$\frac{\partial \; \mathbf{curl\; v}}{\partial t} = \mathbf{curl}(\mathbf{v} \times \mathbf{curl\; v}) \qquad (1.17)$$

The quantity **curl v** is called the vorticity

$$\omega = \mathbf{curl\; v} \qquad (1.18)$$

The vorticity is a vectorial quantity which informs about the local rotational character of the vector field **v**.

In particular, if a flow initially $(t = 0)$ shows no vorticity $(\omega = \mathbf{0})$ then, by virtue of eq 1.17, the vorticity will remain zero subsequently. In other words, if the flow of an inviscid fluid is initially irrotational, *i.e.* no vortices exist in the fluid, then no vortex can be created at any time.

A closer analysis of this problem would reveal more complex situations, such as a fluid flowing around a solid body. In that case the streamlines may take off behind the body at the limit of zero viscosity of the fluid. However, all fluids exhibit some viscosity and no such phenomenon can be observed. Experiments show that vorticity is generally generated in a thin boundary layer, close to a solid surface. It is propagated from the wall by both viscous diffusion and convection. The vortices are transported with the fluid; they are observable for some time after their appearance. If the experiment is made with a circular cylinder moving at a constant velocity, the eddies appear in the wake of the body and their regular distribution constitutes the famous, as well as beautiful, Karman "vortex street".

1.4 Viscous incompressible fluids

1.4.1 Preliminary remarks

If fluids were moving freely in the space one would not have to consider expressly the influence of their viscosity. However, in most natural and experimental situations the fluids are bound by solid surfaces, close to which their velocity is much less than in the bulk of the fluid. In fact, the requirement of zero velocity near to a solid wall has been confirmed both by direct observation and the correctness of theoretical approaches. This yields the no-slip condition

$$\mathbf{v} = \mathbf{0} \qquad (1.19)$$

close to a solid wall. The no-slip condition is a boundary condition for the flow equations of a real fluid. As shown previously, vorticity is created in a boundary layer close to the wall. As far as eddies are confined to that region the bulk flow will not be affected. If more drastic conditions are applied to the system, the whole flow may be perturbed by a turbulent motion generated at the wall. This phenomenon is generally the origin of increased resistances to the motion, e.g. for flying objects in the air or for liquids in pipes. However, some exceptions exist for which the turbulence created reduces the drag on the object, as is the case for golf balls in a certain range of speed.

A flow, or a part of it, is said to be laminar if it is not perturbed by turbulent motions. The particularity of a laminar flow is to appear regular and ordered whereas a turbulent flow is intermittent and disordered. Laminar flows usually exhibit vortex-free patterns that can be made visible by optical interference methods or, more easily, by placing particles of visible size in the liquid.

1.4.2 Microscopic origin of viscosity

The phenomenon of viscosity is related to the transport of momentum between contiguous layers of fluid, moving at different velocities.

In order to illustrate this assertion it is convenient to consider a particular motion in which a liquid between two flat parallel plates is moved under the action of one of the plates. This mobile plate moves at a constant velocity in the x direction. For this simple shearing motion the layers of the fluid slide on each other. The fluid has a constant velocity at every altitude z. The velocity has components $[v(z), 0, 0]$ relative to the orthogonal rectilinear axes.

Two kinds of coupling between adjoining layers cause the exchange of momentum between them: *(i)* intermolecular forces between the particles are the origin of a direct tangential stress; *(ii)* thermal agitation brings particles of higher velocity from the upper to the lower layer, resulting in a force acting on the layers.

For a gas, only the latter effect should be expected to be important. Although we are rather concerned here with the case of liquids, the consideration of gases is very instructive so as to highlight the microscopic origin of viscosity. Besides, gases and liquids are very similar in that their velocity distributions are both Maxwellian at equilibrium (local equilibrium in a layer).

So, let us denote by L the length of the mean free path of the molecules in a gas phase, *i.e.* the mean distance of particles between two successive collisions, by n the number density of the particles and m their mass. The problem is to compute the total momentum passing from a layer at altitude z to the layer below during an interval of time Δt through the interface area S between the layers (Figure 1.1).

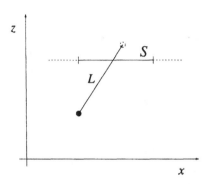

Figure 1.1: Transfer of momentum between contiguous layers of fluid.

The assumption is made that the particles at distances of the order of L from the interface carry their horizontal momentum into the contiguous layer, before they experience a new collision with the surrounding particles, by which they transfer their excess of momentum. Then the horizontal momentum transfer downwards in a time Δt is

$$p_- = \alpha \left[n(kT/m)^{1/2} S \Delta t \right] \left[m\, v(z + 2L/3) \right] \tag{1.20}$$

In eq 1.20 the coefficient α can be calculated from the kinetic theory of gases. The first term in brackets is an estimate of the number of particles crossing the interface S during the interval of time Δt: it is known from statistical mechanics that the mean kinetic energy of a particle is $kT/2$ per degree of freedom; hence $(kT/m)^{1/2}$ is an estimate for the mean vertical velocity of particles. In the second term in brackets the mean position of the particles coming downwards is taken to be $z + 2L/3$: taking into account the probability $P(r) = \exp(-r/L)$ that a molecule has experienced no collision after a distance r, it can be shown that the mean distance to the surface S of the particles which cross the interface is $2L/3$.

In the same way the momentum transported upwards is

$$p_+ = \alpha \left[n(kT/m)^{1/2}S\Delta t\right] \left[m\,v(z - 2L/3)\right]$$

then the net amount of horizontal momentum transferred vertically is

$$\Delta p = p_+ - p_- = \alpha \left[n(kT/m)^{1/2}mS\Delta t\right] \left[m(4L/3)\frac{dv}{dz}\right] \tag{1.21}$$

and the force per unit of surface, σ, exerted on a layer is therefore given by

$$\sigma = -\alpha n(kT/m)^{1/2}m(4L/3)\frac{dv}{dz} \tag{1.22}$$

which is usually called the stress exerted across an element of surface. In the present case it is applied in only one direction. The comparison with Newton's equation of internal friction

$$\sigma = -\eta\,\frac{dv}{dz} \tag{1.23}$$

yields the viscosity

$$\eta = \alpha n(mkT)^{1/2}(4L/3) \tag{1.24}$$

or, introducing the mean velocity $\bar{v}$ and taking for α the value yielded by the kinetec theory of gases

$$\bar{v} = (8kT/\pi m)^{1/2} \quad \text{and} \quad \alpha = (2\pi)^{1/2}$$

leads to

$$\eta = \frac{1}{3}\,n\,m\,\bar{v}\,L \tag{1.25}$$

This law was first demonstrated, and then verified experimentally, by Maxwell around 1860. It constituted one of the first great successes of statistical mechanics.

In particular it allowed description of two unexpected and remarkable properties, that we consider now. For a perfect gas the mean free path is approximately related to the radius R of the gas molecules by

$$n\left(\pi R^2 L\right) \simeq 1 \tag{1.26}$$

expressing the fact that a molecule experiences one collision in the volume $\pi R^2 L$. Therefore, using eq 1.24, one gets

$$\eta \propto \frac{1}{R^2}(mkT)^{1/2} \tag{1.27}$$

in which the symbol $\propto$ means "proportional to". This relation thus leads to the two above-mentioned consequences that (i) the viscosity should be independent of the number density n and (ii) in contrast with liquids, the viscosity increases with temperature.

Moreover, this theory can give reasonable values for real gases, as shown now. First, a more rigorous calculation of the mean free path, taking into account the relative motion of the molecules, leads to

$$4\sqrt{2}\, n\,(\pi R^2 L) = 1 \tag{1.28}$$

Therefore, by substituting this relation into eq 1.24, one finds

$$\eta = (mkT/\pi)^{1/2}/(6\pi R^2) \tag{1.29}$$

An estimate of η for oxygen may illustrate the validity of this molecular theory: eq 1.29 yields $\eta = 2.08 \times 10^{-4}$ g cm^{-1} s^{-1} at 25 °C , whereas the experimental value is 2.05×10^{-4} g cm^{-1} s^{-1}.

1.4.3   Equation of motion of a viscous liquid

In a layer of thickness dz, much larger than the mean free path L, and an area dS in the direction of the flow, the force applied to an element of volume of the fluid can be simply expressed by the difference of the viscous stresses, eq 1.23, on each side of the layer

$$dF = [\eta\frac{dv}{dz}(z + dz) - \eta\frac{dv}{dz}(z)]\, dS$$

Then the force exerted in the x direction per unit volume is

$$\frac{dF}{dV} = \eta\,\frac{d^2v}{dz^2} \tag{1.30}$$

In the general case of a velocity field $\mathbf{v}(\mathbf{r}, t)$ it might be expected that the force exerted on an element of fluid can be expressed with the help of a vector operator of the gradient type. Equation 1.30 suggests that this operator is the Laplace operator Δ, yielding the vectorial identity

$$\Delta\mathbf{v} = \mathbf{i}\,\Delta v_x + \mathbf{j}\,\Delta v_y + \mathbf{k}\,\Delta v_z$$

and that the force per unit volume is

$$\mathbf{f} = \frac{d\mathbf{F}}{dV} = \eta\,\Delta\mathbf{v} \tag{1.31}$$

Then the equation of motion of a viscous liquid is obtained from Euler's equation 1.13 for inviscid liquids by adding $\mathbf{f}$ to yield

$$\partial_t\mathbf{v} + (\mathbf{v}.\nabla)\mathbf{v} = -\nabla(p/\rho) + \mathbf{g} + (\eta/\rho)\Delta\mathbf{v} \tag{1.32}$$

Equation 1.32 was first established by Navier in 1822 and by Poisson in 1829, and later by Saint-Venant in 1843 and by Stokes in 1845. This relation is nowadays known as the Navier-Stokes equation. The parameter η is called the dynamic viscosity. It is often expressed in poise [g cm^{-1} s^{-1}] in the cgs system: the viscosity of water is *ca.* 10^{-2} Poise at 20°C.

In eq 1.32 appears a new quantity, the ratio η/ρ, called the kinematic viscosity ν

$$\nu = \eta/\rho \tag{1.33}$$

The kinematic viscosity is a diffusivity for the velocity $\mathbf{v}$, in analogy to the diffusivity of matter which appears in Fick's law and, like a diffusion coefficient, it has the dimension of m^{2}s^{-1}. The name of "dynamic viscosity" for η reminds that this is the physical parameter used to express the force acting on an element of fluid.

A glance to eq 1.32 shows that ν can be the relevant quantity to describe the flow if no pressure gradient and no volume force (such as gravity) is applied. This conclusion is entirely true in fact only if moreover no boundary condition is imposed involving an applied stress. This happens for instance when a fluid in motion goes back freely to rest (the only boundary condition is then the no-slip condition).

The dynamic viscosity η is the relevant parameter in stationary flows where the velocity varies perpendicularly to the velocity vector. Then both terms of the left hand side of eq 1.32 vanish. A typical example is the stationary flow of a liquid through a circular pipe of constant cross section. The velocity ef the fluid at a distance r from the axis of the pipe is given by Poiseuille formula

$$v(r) = \Delta p\,(R^2 - r^2)/(4\eta l) \tag{1.34}$$

where $\Delta p/l$ is the pressure gradient applied to move the fluid and R is the radius of the pipe.

1.4.4 Dynamical similarity: the Reynolds number

Nowadays wind-tunnel experiments are currently performed by airplane or car engineers to test the aerodynamicity of their machines. The physical principle of these experiments is the dynamical similarity of flows.

For the present discussion it is sufficient to use the Navier-Stokes equation 1.32 in a reduced form by neglecting the contributions of the volume forces

$$\partial_t \mathbf{v} + (\mathbf{v}.\nabla)\mathbf{v} = -\nabla(p/\rho) + (\eta/\rho)\Delta\mathbf{v} \qquad (1.35)$$

which can be transformed with the help of the dimensionless variables

$$v' = v/U \quad r' = r/L \quad t' = tU/L \quad p' = p/(\rho U^2)$$

where L is a characteristic length and U is a characteristic velocity. If, for example, the motion of a rigid sphere in a liquid is considered, L may be the diameter of the sphere and U its speed.

With the help of these new variables, the reduced Navier-Stokes equation is transformed to yield

$$\partial_{t'} \mathbf{v}' + (\mathbf{v}'.\nabla')\mathbf{v}' = -\nabla'p' + \frac{1}{\mathrm{Re}}\Delta'\mathbf{v}' \qquad (1.36)$$

where the superscript ' indicates differential operators corresponding to the variable $r' = r/L$, and Re is the Reynolds number defined by

$$\mathrm{Re} = UL/\nu \qquad (1.37)$$

Re is a dimensionless parameter. Its definition constitutes a powerful tool for the transfer of information from experiments performed at the laboratory scale on various hydrodynamic phenomena to very large (e.g. airplanes) or very small (particles) scales. Via eq 1.36 Re also provides the possibilty of further simplification of the Navier-Stokes equation if the value of Re is very large or very small.

The discussion of electrolyte solutions requires the estimation of the Reynolds number for the particular case where L is of the order of the mean diameter of the particles, *i.e.* 0.1 nm. All liquids commonly used as solvents show dynamic viscosities of the order of 1 cPoise and densities of the order of 1 g cm^{-3}. Then the order of magnitude of Re can be evaluated if for U an estimate of the hydrodynamic velocity of the sphere in the liquid can be made. The order of magnitude of U can be derived from the linear transport theory, where the motion of a particle in a liquid is described at a local level by the action of a friction force $\mathbf{F}$ (eq 1.1). In the steady state of motion this force is supposed to equilibrate the "thermodynamic force"

$$\mathbf{F}_{appl} = -\nabla\mu$$

derived from the chemical potential per particle

$$\mu = \mu^0 + kT \ln(yC)$$

Equation 1.1 may be rewritten in the form

$$\mathbf{v} = -\mathbf{F}/\zeta = \mathbf{F}_{appl}/\zeta \qquad (1.38)$$

for a particle of microscopic size, with ζ the friction coefficient, to yield the stationary velocity of the particle. Thus, we have that

$$\mathbf{F}_{appl} = -\frac{kT}{C}\nabla C \qquad (1.39)$$

if we assume an ideal solution where the activity coefficient y is unity. Making use of Einstein's relation

$$D = kT/\zeta$$

where D is the diffusion coefficient of the particle, we obtain

$$\mathbf{v} = \frac{D}{kT}\mathbf{F}_{appl} = -\frac{D}{C}\nabla C \qquad (1.40)$$

The estimation of $\mathcal{O}(\mathbf{F})$ and $\mathcal{O}(\mathbf{v})$ may be based on the assumption that, typically, $\nabla C \sim C/l$, with $l = 1$ cm and D is of the order of $10^{-5}\text{cm}^2 \text{ s}^{-1}$. Then follows a hydrodynamic velocity of the order of 10^{-5}cm s^{-1} and a Reynolds number of

$$\mathcal{O}(\text{Re}) = 10^{-11}$$

Such very small Reynolds numbers are typical for the motion of particles, ions and molecules, in a liquid, even under the action of an external field (e.g. in an electrophoresis experiment). They permit to simplify the Navier-Stokes equation by neglecting the non-linear convective term $(\mathbf{v}.\nabla)\mathbf{v}$. The resulting relation is known as the Stokes approximation.

1.5 The Stokes approximation

1.5.1 Flow due to a moving sphere at small Reynolds numbers

At low Reynolds numbers the Navier-Stokes equation for a stationary flow, $\partial \mathbf{v}/\partial t \equiv 0$, is

$$\nabla p = \eta \Delta \mathbf{v} \qquad (1.41)$$

if moreover $\mathbf{g}=0$. The application of the **curl** operator to eq 1.41 entails disappearance of the term ∇p

$$\Delta \, \mathbf{curl} \, \mathbf{v} = 0 \qquad (1.42)$$

This equation and the continuity equation for an incompressible liquid, eq 1.6, determine completely the flow of the liquid.

The vector field $\mathbf{v}(\mathbf{r}, t)$ is free of sources meaning that it can be discussed with the help of its accompanying rotational field, defined by a vector $\mathbf{A}(\mathbf{r}, t)$ related to $\mathbf{v}(\mathbf{r}, t)$

$$\mathbf{v} = \mathbf{curl}\,\mathbf{A} \tag{1.43}$$

Here $\mathbf{A}$ is an axial vector (or pseudovector) whereas $\mathbf{v}$ is a polar vector. Axial vectors show the particularity of being changed to their opposites upon inversion of the coordinate system, in contrast to polar vectors (this is the case of the magnetic field in Electromagnetism).

1.5.2   Velocity field around a sphere

Due to the symmetry of a moving sphere of radius R, the velocity $\mathbf{v}$ of the fluid only depends on the distance variable $\mathbf{r}$ (the origin of the coordinate system is taken at the center of the sphere) and on the velocity of the sphere $\mathbf{U}$, and so does the vector $\mathbf{A}$. The only axial vector that can be obtained with $\mathbf{r}$ and $\mathbf{U}$ is the vectorial product $\mathbf{r} \times \mathbf{U}$. Therefore $\mathbf{A}$ must be of the form

$$\mathbf{A} = g(r)\,\mathbf{n} \times \mathbf{U} \quad \text{with} \quad \mathbf{n} = \mathbf{r}/r \tag{1.44}$$

where $g(r)$ is a function of r and $\mathbf{n}$ is the unit vector along r.

The quantity $g(r)\mathbf{n}$ also can be written in the form of $\nabla f(r)$. Then $g(r)$ is the derivative of $f(r)$ with respect to r and $\mathbf{curl}\,\mathbf{A}$ in eq 1.43 can be expressed more simply with help of the relation

$$[\nabla f(r)] \times \mathbf{U} = \mathbf{curl}\,[f(r)\mathbf{U}] \tag{1.45}$$

which holds because $\mathbf{U}$ is constant. Then relation 1.43 is transformed to

$$\mathbf{v} = \mathbf{curl}\,(\mathbf{curl}\,[f(r)\mathbf{U}]) \tag{1.46}$$

The velocity $\mathbf{v}$ satisfies eq 1.42 when it is given by eq 1.46

$$\mathbf{curl}\,\mathbf{v} = \mathbf{curl}\,(\mathbf{curl}\,(\mathbf{curl}\,[f(r)\mathbf{U}])) = (\nabla\,\mathrm{div} - \Delta)(\mathbf{curl}\,[f(r)\mathbf{U}])$$

$$= -\Delta\,\mathbf{curl}\,[f(r)\mathbf{U}] \tag{1.47}$$

for $\mathrm{div}(\mathbf{curl}\,\mathbf{G}) = 0$ for any vector $\mathbf{G}$.

Thus expression 1.42 may be transformed to

$$\Delta^2 \, \mathbf{curl} \, [f(r)\mathbf{U}] = 0 \tag{1.48}$$

The use of eq 1.45 where $\mathbf{U}$ is the constant particle velocity and $f(r)$ is a function only of r permits the reduction of eq 1.48 to

$$\Delta^2[\nabla \, f(r)] = \nabla \, [\Delta^2 f(r)] = \mathbf{0} \tag{1.49}$$

After some algebra in spherical coordinates, integration of this equation and taking into account the fact that $v = 0$ at large distance leads to

$$f(r) = ar + \frac{b}{r} \tag{1.50}$$

Substituting this expression into eq 1.46 and using the no-slip condition on the sphere leads to the components of $\mathbf{v}$

$$v_r = U \cos \theta \, [\frac{3}{2} \frac{R}{r} - \frac{1}{2}(\frac{R}{r})^3] \tag{1.51}$$

$$v_\theta = -U \sin \theta \, [\frac{3}{4} \frac{R}{r} + \frac{1}{4}(\frac{R}{r})^3] \tag{1.52}$$

$$v_\phi = 0 \tag{1.53}$$

with the velocity $\mathbf{U}$ directed along the axis $\theta = 0$ (Figure 1.2).

The relations 1.51 and 1.52 can be advantageously expressed in intrinsic coordinates $(\mathbf{k},\mathbf{n})$ with $\mathbf{k}=\mathbf{U}/U$ and $\mathbf{n}$ as defined by eq 1.44 (then $\mathbf{k}.\mathbf{n} = \cos \theta$), as

$$\frac{\mathbf{v}}{U} = \phi(r)\,\mathbf{k} + \psi(r)\,(\mathbf{k}.\mathbf{n})\,\mathbf{n} \tag{1.54}$$

with

$$\phi(r) = \frac{3}{4} \frac{R}{r} + \frac{1}{4} \, (\frac{R}{r})^3 \quad ; \quad \psi(r) = \frac{3}{4} \frac{R}{r} - \frac{3}{4} \, (\frac{R}{r})^3 \tag{1.55}$$

Eqs 1.54, 1.55 (and r>R) show that $\mathbf{v}.\mathbf{k}> 0$, meaning that the fluid is everywhere moved in the direction of $\mathbf{U}$.

It is convenient to denote by $\mathbf{h}$ the vector

$$\mathbf{h} = (\mathbf{k}.\mathbf{n})\,\mathbf{n} \tag{1.56}$$

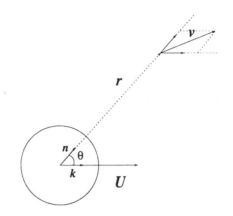

Figure 1.2: Flow around a moving sphere.

The scalar product **k.n** equals the sum of $k_j n_j$ in an arbitrary Cartesian coordinate system for which the components of **k** are k_j and those of **n** are n_j. It is a common convention to write **k.n**$= k_j n_j$, meaning that if an index appears twice in a product, then the sum is performed implicitly with respect to that index, here subscript j. Then the vector **h** can be written alternatively by expressing the components h_1 of **h**

$$h_1 = (k_j n_j) n_1 = (n_j n_1) k_j$$

or

$$h_1 = N_{j1}\, k_j \tag{1.57}$$

in which the sum is performed on the index j, with

$$N_{j1} = n_j n_1 \tag{1.58}$$

The components of the matrix N_{j1} given by eq 1.58 define a tensor of second-rank $\tilde{N}$ which is the dyadic product **n** $\otimes$ **n**

$$\tilde{N} = \mathbf{n} \otimes \mathbf{n} = \frac{\mathbf{r} \otimes \mathbf{r}}{r^2} \tag{1.59}$$

Using these notations the vector **h** may be written in the form

$$\mathbf{h} = \tilde{N}\,\mathbf{k} \tag{1.60}$$

and eq 1.54 takes the form

$$\frac{\mathbf{v}}{U} = \phi(r)\,\mathbf{k} + \psi(r)\,\tilde{N}\,\mathbf{k} \tag{1.61}$$

Equation 1.61 describes the flow of the liquid around the moving sphere, the medium being at rest at infinity. Since $\mathbf{U} = U\mathbf{k}$, eq 1.61 can be rewritten as

$$\mathbf{v} = \tilde{A}\,\mathbf{U} \tag{1.62}$$

where

$$\tilde{A} = \phi(r)\,\tilde{I} + \psi(r)\,\tilde{N} \tag{1.63}$$

with $\tilde{I}$ the second-rank unit tensor.

1.6   Interaction between moving spheres

Two spheres, S_a and S_b, are supposed to move at velocities $\mathbf{v}_a$ and $\mathbf{v}_b$ with respect to the coordinate system in which the fluid is at rest at infinity. The spheres are located at time t at positions $\mathbf{r}_a$ and $\mathbf{r}_b$, respectively. The situation is depicted on Figure 1.3.

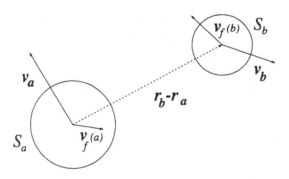

Figure 1.3: Interaction between two moving spheres.

Then the velocity of the medium at $\mathbf{r}_b$, is not simply given by an equation of the type of eq 1.63 because now the fluid is moving with respect to the coordinate system at $\mathbf{r}_a$ as a consequence of the motion of S_b. Let us denote by $\mathbf{v}_f(r)$ the velocity of the fluid at position $\mathbf{r}$ and $\mathbf{v}_r(a)$ the relative velocity of S_a with respect to the fluid at $\mathbf{r}_a$. The two spheres are not necessarily of equal radii; R_a is the radius of S_a and R_b that of S_b. Hence the actions of the spheres on the fluid are different. Different tensors $\tilde{A}$ (eq 1.63) must be considered for S_a and S_b.

$$\tilde{A}_{a \to b} = \phi_a(r)\,\tilde{I} + \psi_a(r)\,\tilde{N} \tag{1.64}$$

where $\phi_a(r)$ and $\psi_a(r)$ stand for the functions defined by eq 1.55 if $R = R_a$; $\tilde{N}$ is given by eq 1.59 where $\mathbf{n}$ must be replaced by $\mathbf{n}_{ab}$, the unit vector in the direction of $\mathbf{r}_b - \mathbf{r}_a$. Then eq 1.64 reproduces the hydrodynamic action of S_a on S_b. Vice versa,

$$\tilde{A}_{b\to a} = \phi_b(r)\,\tilde{I} + \psi_b(r)\,\tilde{N} \tag{1.65}$$

reproduces the hydrodynamic action of S_b on S_a. The same tensor $\tilde{N}$ is used. The reason is given by its definition, eq 1.59.

The velocity of the sphere S_b with respect to the fluid, $i.e.$ the velocity $\mathbf{v}_r(b)$, modified by the 'wake-effect' can be calculated with the help of the simple following model: the relative velocity of a sphere is supposed to be the difference between the velocity expressed in the coordinate system in which the fluid is at rest at infinity and the velocity that the fluid would have if the sphere were not existent. With regard to the sphere S_b follows

$$\mathbf{v}_r(b) = \mathbf{v}_b - \mathbf{v}_f(b) \tag{1.66}$$

This approximation holds the better, the larger the distance between the two spheres. When the distance between the spheres is very large in comparison with their radii, the velocity field of the fluid in the vicinity of point $\mathbf{r}_b$, in the absence of the sphere S_b, is approximately uniform.

Then the sphere moves at that position as if the liquid had a constant velocity $\mathbf{v}_f(b)$ at infinity. According to this model eq 1.62 is modified to

$$\mathbf{v}_f(b) = \tilde{A}_{a\to b}\,\mathbf{v}_r(a) \tag{1.67}$$

and $\mathbf{v}_f(a)$ is given by an analogous equation

$$\mathbf{v}_f(a) = \tilde{A}_{b\to a}\,\mathbf{v}_r(b) \tag{1.68}$$

where

$$\mathbf{v}_r(a) \equiv \mathbf{v}_a - \mathbf{v}_f(a) \tag{1.69}$$

The combination of eqs 1.66-1.69 yields

$$\mathbf{v}_f(b) = \tilde{A}_{a\to b}\left[\mathbf{v}_a - \tilde{A}_{b\to a}(\mathbf{v}_b - \mathbf{v}_f(b))\right] \tag{1.70}$$

or, after rearrangement of the terms

$$\mathbf{v}_f(b) = (\tilde{I} - \tilde{A}_{a\to b}\,\tilde{A}_{b\to a})^{-1}\,\tilde{A}_{a\to b}\,(\mathbf{v}_a - \tilde{A}_{b\to a}\,\mathbf{v}_b) \tag{1.71}$$

Eq 1.71 permits the development of a complete theory of the Oseen tensor. However, in the framework of this study it is sufficient to keep only the first order approximation and neglect the higher order terms. So,

- the products $\tilde{A}_{a \to b}\, \tilde{A}_{b \to a}$ and $\tilde{A}_{b \to a}\, \mathbf{v}_b$ are terms of higher order in R_a and R_b
- the first order terms of $\tilde{A}_{a \to b}$ in R_a and R_b are based on the approximations

$$\phi(r_{ij}) \;=\; \psi(r_{ij}) \;=\; \frac{3}{4}\frac{R_i}{r_{ij}}$$

yielding

$$\tilde{A}_{a \to b} = \frac{3}{4}\frac{R_a}{r_{ab}}(\tilde{I} + \tilde{N}_{ab})$$

and

$$\mathbf{v}_f(b) = \frac{3}{4}\frac{R_a}{r_{ab}}(\tilde{I} + \mathbf{n}_{ab} \otimes \mathbf{n}_{ab})\, \mathbf{v}_a \tag{1.72}$$

Setting

$$\zeta_i = 6\pi\eta R_i \tag{1.73}$$

(see eq 1.1) permits to write eq 1.72 in the form

$$v_f(b) \;=\; \zeta_a\, \tilde{T}_{ab}\, \mathbf{v}_a \tag{1.74}$$

where

$$\tilde{T}_{ab} \;=\; \frac{1}{8\pi\eta r_{ab}}\,(\tilde{I} + \mathbf{n}_{ab} \otimes \mathbf{n}_{ab}) \tag{1.75}$$

is the Oseen tensor.

In presence of the sphere S_a the liquid exerts on the sphere S_b the friction force

$$\mathbf{F}_b = -\zeta_b\,[\mathbf{v}_b - \mathbf{v}_f(b)] \tag{1.76}$$

whereas the absence of the sphere S_a yields a force $-\zeta_b \mathbf{v}_b$

Then

$$\Delta \mathbf{F}_b = \zeta_b\, \Delta \mathbf{v}_b; \qquad \Delta \mathbf{v}_b \;=\; \mathbf{v}_f(b) \tag{1.77}$$

is the additional force acting on the moving sphere S_b produced by the presence of the moving sphere S_a. A suitable expression for $\Delta \mathbf{F}_b$ is obtained from Eqs. 1.74 and 1.77.

It should be emphasized that the name "Oseen tensor", in chemical physics attributed commonly to the tensor expressed by eq 1.75 is somewhat misleading. In hydrodynamics, the Oseen approximation refers to a second-level approximation, see Eqs. 1.71 and subsequent remarks on the approximations. Oseen introduced his second-level approximation in 1910 to get a reliable description of the velocity field at distances greater than R/Re, where R is the radius of the flow. However, since the Reynolds numbers for the systems investigated in solution chemistry generally are particularly low, it follows that the first-level description presented here is entirely satisfying.

For a more detailed description of this topic the reader can be referred to Refs. [4-8].

Bibliography

[1] G. K. Batchelor *An Introduction to Fluid Dynamics*, Cambridge University Press, CambridgeUK, 1967.

[2] V. G. Levich *Physicochemical Hydrodynamics*, Prentice Hall, Englewood Cliffs, New Jersey 1962.

[3] L. D. Landau and E. M. Lifschitz *Mécanique des fluides*, Edition Mir, Moscou, 1971.

[4] P. Résibois et N. Hassele-Schuermans in: I. Prigogine (ed.), *Advances in Chemical Physics*, Vol. XI, Intersciene, New York 1967.

[5] P. Résibois *Electrolyte Theory: An Elementary Introduction to a Microscopic Approach*, Harder and Row, New York, 1968.

[6] H.L. Friedman *Physica* 30 (1964) 537.

[7] H.L. Friedman *J. Chem. Phys.* 42 (1964) 450.

[8] H.L. Friedman *J. Chem. Phys.* 42 (1964) 459.

Chapter 2

Electrostatics

2.1   Introduction

The aim of this paragraph is to recall the basic principles of electrostatics and their application to electrolytes and interfaces. After presenting the quasi static electric fields, we discuss some dielectric properties. We emphasize then the Poisson-Boltzmann equation and its solutions for different symmetries, giving the Gouy-Chapman, the Debye-Hückel and the Lifson-Katchalsky approximations.

In this chapter we follow the treatment given by Russell, Saville and Schowalter [8].

2.2   Electrostatic fields

We begin with Maxwell's equations, simplified for electrostatics (see Feynman, Leighton and Sands 1964 [1]):

$$\nabla \cdot \mathbf{E} = \frac{\rho^{(e)}}{\varepsilon_o} \qquad (2.1)$$

and

$$\nabla \times \mathbf{E} = 0 \tag{2.2}$$

The fundamental entities introduced here are the electric field $\mathbf{E}$, the total electric charge per unit volume $\rho^{(e)}$, and ε_o the primitivity of vacuum. Combining 2.1 with the divergence theorem, relates the electric field on a closed surface S, to the charge Q, enclosed in the volume V, as

$$\int_S \mathbf{E} \cdot \mathbf{n} dS = \frac{1}{\varepsilon_o} \int_V \rho^{(e)} dV = \frac{Q}{\varepsilon_o} \tag{2.3}$$

Here $\mathbf{n}$ is the outer unit normal. Consequently, for a spherical surface of radius r centered on a point charge in vacuum,

$$\mathbf{E} \cdot \mathbf{n} = \frac{Q}{4\pi\varepsilon_o r^2} \tag{2.4}$$

indicating that

$$\mathbf{E} = \frac{Q}{4\pi\varepsilon_o r^3}\mathbf{r} \tag{2.5}$$

where $\mathbf{r}$ is centered on the point charge. Since the electric field is defined as the force per unit charge, the force exerted by one point charge on another at relative position $\mathbf{r}_{12}$ (Coulomb's law) is

$$\mathbf{F}_{12} = \mathbf{E}_1 Q_2 = \frac{Q_1 Q_2}{4\pi\varepsilon_o r_{12}^3}\mathbf{r}_{12} \tag{2.6}$$

Note that SI system is used here.

According to equation 2.2, the electric field is conservative, *i.e.* the line integral of $\mathbf{E} \cdot \mathbf{t}$ (where $\mathbf{t}$ is tangent to a closed curve) is zero. Thus, there exists a potential function such that

$$\mathbf{E} = -\nabla\psi \tag{2.7}$$

The potential must satisfy (see equation 2.1)

$$\nabla^2 \psi = -\frac{\rho^{(e)}}{\varepsilon_o} \qquad (2.8)$$

For example, around an isolated point charge Q, in three dimensions,

$$\frac{1}{r^2}\frac{\partial}{\partial r}r^2\frac{\partial \psi}{\partial r} = 0 \qquad (2.9)$$

with $\psi \to 0$ when $r \to \infty$
and

$$\lim_{r\to 0}\int \frac{\partial \psi}{\partial r}dS = -\frac{Q}{\varepsilon_o} \qquad (2.10)$$

The solution for the potential

$$\psi = \frac{Q}{4\pi\varepsilon_o}r^{-1} \qquad (2.11)$$

corresponds to the electric field of equation 2.3.

The effect of matter on the electrostatic field can be described in an empirical manner by the use of a dielectric constant ε. Let us consider the particular case of a capacitor containing a dielectric medium. Consider a spherical conductor of radius a. Equation 2.2 connects charge and potential in vacuum, so that the potential of the sphere ψ_s, which is the work to bring in a unit charge from infinity, and the sphere charge Q, are related by

$$Q = 4\pi\varepsilon_o a\psi_s \qquad (2.12)$$

The factor $4\pi\varepsilon_o a$ is the capacity of the sphere.

If the vacuum is replaced by a dielectric medium, the capacity of the sphere increases because of polarization of the dielectric. Electric fields polarize matter in two ways: by orienting molecules with permanent dipoles and by deforming electron clouds of each molecule. The polarization vector $\mathbf{P}$ is related to the characteristics of individual dipoles by the relation:

$$\mathbf{P} = NQ\mathbf{d} \qquad (2.13)$$

N represents the number of dipoles per unit volume, Q is the magnitude of the charge separated to produce the dipole, and $\mathbf{d}$ is a vector describing the average orientation of the dipole and the charge separation distance. In linear materials, polarization from incident field is expressed as

$$\mathbf{P} = N\alpha\varepsilon_o\mathbf{E} \tag{2.14}$$

where α is the polarisability of the molecule, with dimension L^3. The product $N\alpha$ is called the dielectric susceptibility of the material, χ.

The polarization vector is then used to define the volumetric polarization charge density $\rho^{(p)}$ as

$$\nabla \cdot \mathbf{P} = -\rho^{(p)} \tag{2.15}$$

Combining equations 2.1, 2.14 and 2.15 yields

$$\varepsilon_o\nabla \cdot (1 + \chi)\mathbf{E} = \rho^{(e)} - \rho^{(p)} \equiv \rho^{(f)} \tag{2.16}$$

If the free charge density $\rho^{(f)}$ is zero and the dielectric homogeneous, then equations 2.1 and 2.16 together, indicate that the volumetric polarization charge is also zero. Note that, however, a polarization charge will appear at surface.

Now we examine a spherical capacitor immersed in a homogeneous dielectric. If the dielectric has no free charge, then

$$\varepsilon_o\nabla \cdot (1 + \chi)\mathbf{E} = 0 \tag{2.17}$$

and since the field is still irrotational, a potential exists $viz.$

$$\psi = Ar^{-1} \tag{2.18}$$

However the situation at the surface differs from that in vacuum, owing to the polarization charge. From equation 2.15, the divergence theorem applied to the Gaussian surface, defined as the surface enclosing the free charge plus the polarization charge at the surface of the dielectric, we have

$$\mathbf{P}.\mathbf{n} = -\frac{Q^{(p)}}{4\pi a^2} \tag{2.19}$$

where $Q^{(p)}$ is the polarization charge on the surface of the dielectric. Applying equation 2.3 to the Gaussian surface yields

$$\mathbf{E} \cdot \mathbf{n} = \frac{Q + Q^{(p)}}{4\pi\varepsilon_o a^2} \tag{2.20}$$

since the total electric charge consists of the free charge on the capacitor Q, plus the polarization charge on the surface of the dielectric $Q^{(p)}$.
Thus

$$\mathbf{E} \cdot \mathbf{n} = \frac{Q}{4\pi\varepsilon_o a^2} - \chi\mathbf{E} \cdot \mathbf{n} \tag{2.21}$$

This determines the integration constant A and

$$Q = 4\pi\varepsilon_o(1 + \chi)a\psi_s \tag{2.22}$$

Hence the capacity increases by the factor $(1 + \chi)$ compared with the situation in vacuum; $(1+\chi)$ is often designed the dielectric constant ε. Similarly we can show that the electric field at a distance r from a point charge in a uniform dielectric in equation 2.5 and the Coulomb force in equation 2.6 are each divided by a factor ε.

2.3   Boundary conditions

At this point we have a mathematical structure that describes fields in bulk matter. A complementary description of conditions prevailing at interfaces between two materials can be derived by applying the same balance equations to a disc-shaped volume of area πa^2 and height h in each material and to a simple closed curve of height h in each material and side S. Using the divergence theorem and the disc-shaped volume with equation 2.1, we have in the limit as $h \rightarrow 0$

$$\varepsilon_o[(\mathbf{E} \cdot \mathbf{n})_1 + (\mathbf{E} \cdot \mathbf{n})_2] = q \tag{2.23}$$

where the subscripts indicate the side of the interface on which the quantity in parenthesis is evaluated, and q stands for the total charge per unit area. From the expression for the polarization equation 2.15

$$(\mathbf{P} \cdot \mathbf{n})_1 + (\mathbf{P} \cdot \mathbf{n})_2 = -q^{(p)} \tag{2.24}$$

and so

$$[(\varepsilon_o\mathbf{E} + \mathbf{P}) \cdot \mathbf{n}]_1 + [(\varepsilon_o\mathbf{E} + \mathbf{P}) \cdot \mathbf{n}]_2 = q^{(f)} \tag{2.25}$$

Here $q^{(f)}$ stands for the free charge $q - q^{(p)}$, beyond that due to polarization, i.e., charge positioned at the interface by means other than polarization. Thus for linear dielectrics

$$[\varepsilon_o(1 + \chi)\mathbf{E} \cdot \mathbf{n}]_1 + [\varepsilon_o(1 + \chi)\mathbf{E} \cdot \mathbf{n}]_2 = q^{(f)} \tag{2.26}$$

For the system defined above, equation 2.26 yields

$$Charge \simeq \varepsilon_o[(\mathbf{E} \cdot \mathbf{n})_2\pi a^2 + (\mathbf{E} \cdot \mathbf{n})_1\pi a^2 + (\mathbf{E} \cdot \mathbf{n})_h 2\pi ah] \tag{2.27}$$

where the subscript h indicates the lateral side of the cylindrical surface under consideration.

If only one of the materials is a conductor a current normal to the surface must be zero so the corresponding electrostatic field vanishes.

A condition on the tangential component of the field at the interface is obtained by evaluating the line integral of $\mathbf{E} \cdot \mathbf{t}$ around a rectangular path in a plane perpendicular to the surface. Taking the limiting circuit where the path length is perpendicular to the surface shrinks to zero shows using equation 2.2, that

$$(\mathbf{E} \cdot \mathbf{t})_1 + (\mathbf{E} \cdot \mathbf{t})_2 = 0 \tag{2.28}$$

For the particular closed circuit described here we get

$$(\mathbf{E} \cdot \mathbf{t})_u h + (\mathbf{E} \cdot \mathbf{t})_2 S + (\mathbf{E} \cdot \mathbf{t})_d h + (\mathbf{E} \cdot \mathbf{t})_1 S = 0 \tag{2.29}$$

where the subscripts u and h indicates upward and downward orientations respectively.

Here t is any vector tangent to the interface and it follows that the potentials on either side of the interface differ by at most a constant. If no work is done in transferring charge across the interface, the constant is zero.

2.4   Electric stress tensor

The study of interactions between macroscopic bodies requires the force per unit area, *i.e.* the stress that an external field exerts on a surface.

We will see first the expression of the stress tensor in a homogeneous fluid, then illustrate its application to a sphere immersed in an uncharged dielectric.

The derivation of the stress in a fluid dielectric containing free charge is not straightforward, in part because of the difficulty in establishing how the presence of the electric field contributes to the pressure and thereby alters the stress. The derivation below is based on that of Landau and Lifshitz (1960) [2].

First the force due to the electric field acting on an isolated dipole is derived. Consider a pair of charges, Q and $-Q$, at relative position d. The electrical force on the pair is

$$-Q\mathbf{E}(\mathbf{x}) + Q\mathbf{E}(\mathbf{x} + \mathbf{d}) \tag{2.30}$$

Expanding the second term yields

$$-Q\mathbf{E}(\mathbf{x}) + Q\mathbf{E}(\mathbf{x}) + Q\mathbf{d} \cdot \nabla \mathbf{E} + \ldots\ldots\ldots \tag{2.31}$$

and taking the limit $\mathbf{d} \to 0$ with $Q\mathbf{d}$ fixed produces the expression

$$(Q\mathbf{d}) \cdot \nabla \mathbf{E} \tag{2.32}$$

for the force on an individual dipole. For N such dipoles per unit volume the force will be as in equation 2.6

$$\mathbf{P} \cdot \nabla \mathbf{E} \tag{2.33}$$

Accordingly, the force per unit volume acting on the free charge and dipole is

$$\rho^{(f)}\mathbf{E} + \mathbf{P} \cdot \nabla\mathbf{E} \tag{2.34}$$

These are body forces and must be balanced by the pressure gradient ∇p^*
At equilibrium

$$-\nabla p^* + \rho^{(f)}\mathbf{E} + \mathbf{P} \cdot \nabla\mathbf{E} = 0 \tag{2.35}$$

With the expressions relating charge and dipole density to field strength, e-quations 2.14 and 2.15, we can transform this expression into one involving the divergence of a tensor as

$$-\nabla p^* + \nabla \cdot (\varepsilon\varepsilon_o\mathbf{E}\mathbf{E} - \frac{1}{2}\varepsilon_o\mathbf{E} \cdot \mathbf{E}\mathbf{d}) = 0 \tag{2.36}$$

The pressure p^* differs from that in the absence of an electric field owing to electrical modifications to the short-range intermolecular forces. Accordingly, we identify the "pressure" due to kinetic energy and short-range intermolecular forces without electrical effects as p and write

$$p^* = p + \frac{1}{2}\varepsilon_o[\varepsilon - 1 - \rho(\frac{\partial\varepsilon}{\partial\rho})_T]\mathbf{E} \cdot \mathbf{E} \tag{2.37}$$

Here ρ denotes the density of the material
and the derivative $(\partial\varepsilon/\partial\rho)$ is taken at constant temperature T. Now , the divergence of the total stress yields

$$-\nabla p + \nabla \cdot \{\varepsilon\varepsilon_o\mathbf{E}\mathbf{E} - \frac{1}{2}\varepsilon\varepsilon_o[1 - \frac{\rho}{\varepsilon}(\frac{\partial\varepsilon}{\partial\rho})_T]\mathbf{E} \cdot \mathbf{E}\mathbf{d}\} = 0 \tag{2.38}$$

The electric stress tensor reduces to what is known as the Maxwell form for the vacuum where $\varepsilon = 1$. Equation 2.38 also can be written in the form

$$-\nabla[p - \frac{1}{2}\varepsilon_o\rho(\frac{\partial\varepsilon}{\partial\rho})_T\mathbf{E} \cdot \mathbf{E}] - \frac{1}{2}\varepsilon_o\mathbf{E} \cdot \mathbf{E}\nabla\varepsilon + \rho^{(f)}\mathbf{E} = 0 \tag{2.39}$$

to recover the force arising from the action of the field on the local free charge.

To illustrate use of the electric stress tensor, a familiar result is derived: the force on a conducting sphere. A conducting sphere is immersed in a uniform (non-conducting) dielectric in the presence of a uniform field $\mathbf{E}_\infty$. The free charge in the dielectric is zero, while the surface charge found on the conductor is Q. Clearly, the force must emerge as $Q\mathbf{E}_\infty$, since the sphere appears as a point when wieved on a large length scale and Coulomb's law must apply.

From equation 2.17 and the uniformity of the dielectric, we deduce the form of the potential as

$$\psi = -(1 + Ar^{-3})\mathbf{E}_\infty \cdot \mathbf{x} + Br^{-1} \tag{2.40}$$

The field must have the requisite behavior far from the sphere. Since the sphere is a conductor with charge Q, the boundary conditions show that $A = -a^3$ and $B = Q/4\pi\varepsilon\varepsilon_o$.

The electric stress is calculated from the potential but the force on the sphere also includes a contribution from the inhomogeneous pressure generated by the electric field. Integrating equation 2.39 shows that

$$p - \frac{1}{2}\varepsilon_o\rho(\frac{\partial\varepsilon}{\partial\rho})_T\mathbf{E} \cdot \mathbf{E} = constant \tag{2.41}$$

Thus the pressure variation due to the field is cancelled by electrical effects and the net force is

$$\mathbf{F} = \int_S \varepsilon\varepsilon_o(\mathbf{E}\mathbf{E} - \frac{1}{2}\mathbf{E} \cdot \mathbf{E}d) \cdot \mathbf{n}r^2 d\Omega \tag{2.42}$$

with $d\Omega = \sin\theta d\theta d\phi$. Use of the divergence theorem converts the integral to one over a spherical surface with an infinitely large radius. Now we need only those parts of this integral that survive the limiting process and the force on the sphere is

$$\mathbf{F} = Q\mathbf{E}_\infty \tag{2.43}$$

As expected, the force is the same as that on a concentrated point charge placed in an undisturbed field.

A final point worth noting concerns the term

$$\frac{1}{2}\varepsilon_o\rho(\frac{\partial\varepsilon}{\partial\rho})_T\mathbf{E}\cdot\mathbf{E} \tag{2.44}$$

For problems involving rigid bodies immersed in incompressible, homogeneous materials, reference to this term can be avoided by absorbing it into a modified pressure.

2.5 The Gouy-Chapman model of the diffuse layer

The investigation of detailed structure of the electrostatic field requires knowledge of ion distribution because the field produces a net charge in the electrolyte adjacent to the interface. Ions whose charge is opposite to the sign of the charge on the interface will be attracted and the others will be repulsed. At the same time, each ion participates in the randomizing thermal motion in solution. It follows that the fluid adjacent to the charged interface contains a charge which balances the surface charge, making the combination of surface and solution electrically neutral. The region containing the surface charge is often called the compact or Stern layer, while the region where ions move freely under the influence of electrical and thermal forces is termed the diffuse layer. Together, these make up the electric double layer. Our task is to describe the structure of this double layer, especially the diffuse region.

According to the earlier development, the electrostatic potential must satisfy

$$\nabla^2\psi = -\frac{1}{\varepsilon\varepsilon_o}\rho^{(f)} \tag{2.45}$$

for a homogeneous linear dielectric, so the first task is to describe the ion concentrations that produce the free charge density $\rho^{(f)}$. Because the ions in the diffuse region are in equilibrium, the force which equals the gradient of the electrochemical potential must vanish as

$$kT\nabla\ln n^k + ez^k\nabla\psi = 0 \tag{2.46}$$

Thus the ions follow the Boltzmann distribution

$$n^k = n_b^k \exp(-ez^k\psi/kT) \qquad (2.47)$$

It should be noted that the potentials appearing in equations 2.45 and 2.47 are, strictly speaking different. The potential in the Boltzmann equation represents the potential of mean force, whereas in the Poisson equation it is the local average potential [1]. Nevertheless, to simplify matter, differences are ignored.

Recognizing that the free-charge density equals the local excess of ionic charge arising from N ionic species, *i.e.*

$$\rho^{(f)} = \sum_1^N ez^k n^k \qquad (2.48)$$

and combining the various expressions leads to the Poisson-Boltzmann equation describing the electrostatic potential in ionic solutions,

$$\Delta\psi = -\frac{1}{\varepsilon\varepsilon_o}e\sum_1^N z^k n_b^k \exp(-ez^k\psi/kT) \qquad (2.49)$$

This equation is the basis of the Gouy-Chapman model of the diffuse charge cloud adjacent to a charged surface. It was discovered by the french physicist Gouy from Lyon in 1910 [4] and rediscovered in 1913 by Chapman [5] (not the same Chapman as for Chapman-Enskog approximation).

The principal assumptions thus far are that the electrolyte is a solution ideal for other aspects, with uniform dielectric properties, the ions are point charges, and the potential of mean force and the average electrostatic potential are identical. A considerable amount of work has been done to identify limitations of this equation. However, more detailed formulations, ranging from analytical modifications to Monte-Carlo calculations, lead to the conclusion that the Poisson-Boltzmann equation provides very accurate results for the conditions of interest here, *i.e.* electrolyte concentrations that do not exceed 1M and surface potential less than 200mV. In this regard see Haydon (1964) [6], Sparnaay (1972) [7], Russel (1989) [8], Hunter (1993) [9] for summaries of other investigations.

---

[1] The potential of mean force is the potential whose gradient gives the average force acting on an ion, whereas the local average potential is the canonical ensemble average of the electrostatic potential (see Mc Quarrie 1976) [3]

Another condition is required to complete the specification of the potential. At the interface, the relation between the charge and the potential can be established from equation 2.45 as

$$\int_S \nabla\psi \cdot \mathbf{n} dS + \frac{Q^{(f)}}{\varepsilon\varepsilon_o} = 0 \tag{2.50}$$

where $Q^{(f)}$ is the free charge enclosed by the surface. If all the charge is on the interface, $Q^{(f)}$ is equal to the surface charge Q, and for a uniform surface

$$\nabla\psi \cdot \mathbf{n} + \frac{q}{\varepsilon\varepsilon_o} = 0 \tag{2.51}$$

where q is the charge density on the surface (charge per unit area).

2.5.1 Diffuse layer near a plate

Next we investigate some solutions of the equations embodied in the Gouy-Chapman model. For a flat interface and a $z-z$ symmetrical electrolyte $e.g.$ KCl, the Poisson-Boltzmann equation becomes

$$\frac{d^2\psi}{dx^2} = \frac{2ez}{\varepsilon\varepsilon_o} n_b \sinh\left(\frac{ez\psi}{kT}\right) \tag{2.52}$$

where z is the valence of the cationic species. Linearization of the sinh function for small dimensionless potentials, $e\psi/kT$ leads to the Debye-Hückel approximation, $i.e.$

$$\frac{d^2\psi}{dx^2} = \frac{2e^2 z^2 n_b}{\varepsilon\varepsilon_o kT}\psi \tag{2.53}$$

The expression $(\varepsilon\varepsilon_o kT/2e^2 z^2 n_b)^{1/2}$ represents the Debye screening length, symbolized by κ^{-1}. The solution of equation 2.53 gives

$$\psi = \psi_s \exp(-\kappa x) \tag{2.54}$$

Where ψ_s represents the potential at the surface $x = 0$. Hence the potential and space charge are non-zero in a region of thickness κ^{-1} adjacent to the interface.

The exact solution of the non linearized Poisson-Boltzmann equation 2.52 can be obtained explicitly under an analytical form, this property being closely related to the one dimensional character of the present problem . By multiplying both sides by $d\psi/dx$ one obtains exact differentials which, after integrating twice and applying the boundary conditions yields

$$\Psi = 2\ln\frac{1 + \exp(-\kappa x)\tanh(\frac{1}{4}\Psi_s)}{1 - \exp(-\kappa x)\tanh(\frac{1}{4}\Psi_s)} \tag{2.55}$$

where $\Psi = ez\psi/kT$. The surface charge follows from equation 2.51 as

$$q = 2(2\varepsilon\varepsilon_o kTn_b)^{1/2}\sinh(\frac{1}{2}\Psi_s) \tag{2.56}$$

From equation 2.55 we recover the Debye-Hückel approximation for $\Psi_s/4 \ll 1$. The factor $\frac{1}{4}$ accounts for the numerical accuracy of the Debye-Hückel formula for dimensionless potentials somewhat larger than unity. For large positive surface potentials $\Psi_s \gg 1$, and $x > 0$,

$$\Psi \simeq 2\ln\frac{1 + \exp(-\kappa x)}{1 - \exp(-\kappa x)} \tag{2.57}$$

leaving the local potential independent of the surface potential away from the interface. For negative potentials, the negative and positive signs in front of the exponentials are reversed. Similarly, for $\kappa x \gg 1$, $\exp(-\kappa x)$ is small and the general solution 2.55 of the Gouy-Chapman equation reduces to

$$\Psi \simeq 4\tanh(\frac{1}{4}\Psi_s)\exp(-\kappa x) \tag{2.58}$$

and so the decay is always exponential far from the surface. Thus, for large surface potentials, there is a saturation effect, and viewed from a large distance, the surface potential is 4.

2.5.2   Diffuse layer around a sphere

We turn now to the spherical geometry, where curvature complicates

matters and precludes analytical solution in the general case. However, approximate solutions depict qualitative features accurately. For a $z-z$ electrolyte we must solve

$$\frac{1}{r^2}\frac{\partial}{\partial r}(r^2\frac{\partial\psi}{\partial r}) = \frac{2eZn_b}{\varepsilon\varepsilon_o}\sinh(\frac{ez\psi}{kT}) \qquad (2.59)$$

subject to $\psi = \psi_s$ at $r = a$ and $\psi \to 0$ at $r \to \infty$

No analytical solution exists, leaving numerical calculations for specific situations or asymptotic solutions to describe limiting behavior.

For thin double layers, $i.e.$ $\kappa a >> 1$, we can rescale the equations with

$$r = a[1 + y/(\kappa a)]$$

$$\Psi = \frac{ez\psi}{kT}$$

to obtain

$$\frac{d^2\Psi}{dy^2} + \frac{2}{(\kappa a)[1 + y/(\kappa a)]}\frac{d\Psi}{dy} = \sinh\Psi \qquad (2.60)$$

with
$\Psi = \Psi_s$ at $y = 0$
$\Psi \to 0$ as $y \to \infty$
For $\kappa a \to \infty$, we recover a description pertinent to a flat interface. Thin double layers, therefore, exhibit a saturation effect similar to that for planar layers, $i.e.$ for $\Psi_s >> 1$

$$\Psi \simeq 4\exp(-y)$$

so the potential decays as an exponential. Loeb, Overbeek and Wiersema (1961) [10] gave the first comprehensive numerical treatment of spherical systems and tabulated extensive results for both symmetrical and unsymmetrical electrolytes (the LOW Tables). The LOW Tables also include numerous comparisons between exact (numerical) results and approximate formulas of one sort or another. For example, the formula

$$q = \frac{Q}{4\pi a^2} = \frac{\varepsilon\varepsilon_o kT}{ez}\kappa[2\sinh(\frac{1}{2}\Psi_s) + \frac{4}{\kappa a}\tanh(\frac{1}{4}\Psi_s)] \qquad (2.61)$$

adapted from LOW Tables, gives the surface charge density to within 5 per cent for $\kappa a > 0.5$ for any surface potential.

The numerical results also demonstrate that saturation effect at high potential is a general feature of spherical double layers. Because the potential diminishes with distance from the surface, linearization becomes appropriate and the decay is exponential, *viz.*

$$\Psi \simeq \Psi_A \frac{a}{r}\exp(-\kappa(r-a)) \qquad (2.62)$$

Numerical results from the LOW Tables show how the saturation potential Ψ_A depends on κa. Many analytical approximations have been developed for the spherical diffuse layer, but the advent of fast, efficient numerical schemes (HNC, hypernetted chain equation), as well as that of the mean spherical approximation (MSA) has diminished their utility (see further parts of this monograph). Moreover MSA and HNC approximations can be applied to the evaluation of most practical transport coefficients of electrolytes in a large variety of experimental situations, as it will be seen in other parts of this book.

It should be noticed that other works in the field of electrolytes have been made recently by Fisher and his group, for equilibrium and static properties. Transport or dynamic properties where not considered by them for the moment.

They revisited the main topics of electrolyte theory [11] [12] [13], giving formulations, including ion association in electrolytes [14] [15] and counterion condensation on polyelectrolytes and charged surfaces [16] [17] [18] [19] [20].

The interested reader should look at those papers which cannot be presented in details in the framework of this introductory chapter. More details on HNC and MSA and their applications to bulk electrolytes can be found in the monograph of Barthel *et al.* [21].

2.5.3   Repulsion between charged plates

To calculate the repulsive force, we need the local electrostatic potential and the local stress. Combining the equilibrium expression 2.39 with the Gouy-Chapman model of the diffuse layer and integrating yields the general relation

$$p + kTb \sum_{1}^{N} n_b^k [1 - \exp(-ez^k \psi/kT)] = p_b \qquad (2.63)$$

From this equation and from that for the total stress cf. equation 2.38, we calculate the repulsive force per unit area, F, on either of two identical parallel plates, immersed in an ionic solution. A force balance is constructed on a system bounded by the midplane (parallel to the two interfaces) and a parallel surface far away. The forces on the system consist of the force on the plate, F, the pressure on the system boundary at infinity, and the force on the midplane, where, because of symmetry, the electric field vanishes. Since the electric stress on the midplane is zero, the force per unit area follows as

$$F = kT \sum_{1}^{N} n_b^k [\exp(-ez^k \psi_o/kT) - 1] \qquad (2.64)$$

Here the potential ψ_o is evaluated in the midplane. The expression in brackets is simply the excess ionic concentration at the midplane, so the repulsive force per unit area is equal to the osmotic pressure.

For low potentials the force is proportional to the square of the potential *viz.*

$$F = \frac{e^2 z^2 n_b}{kT} \psi_0^2. \qquad (2.65)$$

Then, ignoring interactions between the plates and simply adding potentials from two isolated plates equation 2.61 yields

$$F = 64kTn_b \tanh^2(\frac{1}{4}\Psi_s) \exp(-\kappa h) \qquad (2.66)$$

This non-linear superposition approximation clearly requires that the plate separation, h, be large compared with the Debye length.

To obtain accurate values for the variation of the repulsive force with separation, equation 2.52 must be solved. Constant-charge and constant-potential boundary conditions furnish bounds on the force-distance relation for surfaces that regulate their charge according to the mass action equilibria studied in connection with the isolate plate.

For a $z - z$ electrolyte, the Poisson-Boltzmann equation can be integrated once analytically.

A second integration yields

$$\int_{\zeta_0}^{\zeta} (\beta^2 - 1)^{-\frac{1}{2}} (\beta - \zeta_0)^{-\frac{1}{2}} d\beta = \sqrt{2}\kappa x, \tag{2.67}$$

where

$$\zeta = \cosh\,(ez\psi/kT) \tag{2.68}$$

and the subscript indicates conditions at the midplane. This expression can be evaluated numerically with either constant charge or constant potential at the surface $x = h/2$. Then, using the potential at the miplane, the osmotic pressure or repulsive force is calculated.

Note that at small separations the constant charge results begin to diverge, reflecting the singular interaction at contact.

At this point it is important to recognize that interactions between surfaces with unequal charges or potentials differ from those just studied. The force can be calculated in much the same way but the lack of symmetry about the midplane introduces an additional stress,

$$-\frac{1}{2}\varepsilon\varepsilon_0\mathbf{E}\cdot\mathbf{E}. \tag{2.69}$$

The linearized form of the force

$$F = kTn_b\left(\frac{ez\psi}{kT}\right)_0^2 - \frac{1}{2}\varepsilon\varepsilon_0\left(\frac{\partial\psi}{\partial x}\right)_0^2 \tag{2.70}$$

illustrates how the additional stress counteracts the osmotic repulsion. This stress alters the interaction qualitatively as well as quantitatively

All this is illustrated by the solution of the Debye-Hckel equation for a $z - z$ electrolyte between two plates separated by a distance h. The potential is

$$\Psi = \frac{ez\psi}{kT} = A\cosh\kappa y + B\sinh\kappa y, \tag{2.71}$$

and so the force is simply

$$F = kTn_b(A^2 - B^2) \tag{2.72}$$

The constants A and B follow the boundary conditions.

For constant potentials, $i.e.$,

$$\Psi = \left[\begin{array}{ll} \Psi_+, & y = h/2 \\ \Psi_-, & y = -h/2 \end{array} \right. \tag{2.73}$$

we obtain

$$A = \frac{1}{2}\frac{\Psi_+ + \Psi_-}{\cosh\frac{1}{2}\kappa h}, \quad B = \frac{1}{2}\frac{\Psi_+ - \Psi_-}{\sinh\frac{1}{2}\kappa h}. \tag{2.74}$$

Thus, for surfaces at different potentials, repulsive interactions at large separations can change to attraction at small separations if the surface potentials are held constant. The behavior at close separations results from a change in the sign of the charge on one of the plates.

Similarly, for surfaces where charges are held constant, $i.e.$,

$$q = \left[\begin{array}{ll} q_+, & y = h/2, \\ q_-, & y = -h/2, \end{array} \right. \tag{2.75}$$

$$A = \frac{1}{2}\frac{q_+ + q_-}{\sinh\frac{1}{2}\kappa h}, \quad B = \frac{1}{2}\frac{q_+ - q_-}{\cosh\frac{1}{2}\kappa h}. \tag{2.76}$$

Here q_+ and q_- denote surface charges scaled on $\varepsilon\varepsilon_0\, kTk/ez$. These formulas show clearly the singular behavior of the force as two plates with identical charges are brought close together. Furthermore, depending on the relative magnitude of the two charges, the force can change from attraction to repulsion as the plates are brought together. Interactions between particles with dissimilar charges are at the core of the subject of heterocoagulation

2.6   Repulsion between charged spheres

Given the case of single spheres, the absence of closed-form solutions for the two-sphere problem comes as no surprise. Moreover, because of curvature, the repulsive force derives from both osmotic pressure and an electric stress. We can use the general relation, equation 2.63, along with the equilibrium condition, to show that the repulsive force can be obtained from an integration over the central plane as

$$\mathbf{F} = \int_s kT \sum_1^N n_b^k[\exp(-ez^k\psi_0/kT) - 1)\mathbf{n}\,dS + \int_s \varepsilon\varepsilon_0[\mathbf{E}\mathbf{E} - \frac{1}{2}\mathbf{E}\cdot\mathbf{E}\mathbf{d}]\cdot\mathbf{n}\,dS \tag{2.77}$$

A variety of approximations provides insight into the qualitative and quantitative behavior. The Derjaguin approximation, for example, is applicable for separations small compared with the radius of the spheres (Derjaguin, 1934). Under such conditions, elements on each sphere interact as parallel plane elements at the same separation; the total interaction is a sum over the infinitesimal elements. To proceed formally, we adopt a polar cylindrical coordinate system with its axis joining the centers of the spheres of radius a, separated by the distance h and centered at midpoint.

r is the interparticular distance on the midplane and z the distance on the center to center axis.

A sphere surface is defined by

$$z_* = \frac{1}{2}h + a[1 - (1 - r_*^2/a^2)^{1/2}]. \tag{2.78}$$

Scaling distances as

$$z = \kappa z_*, \quad r = (\kappa a)^{1/2} r_*/a \tag{2.79}$$

and expanding the potential as

$$\Psi = \Psi_1(r, z) + (\kappa a)^{-1}\Psi_2(r, z) + \dots, \tag{2.80}$$

leads to

$$\frac{\partial^2 \Psi_1}{\partial z^2} = \sinh \Psi_1 \tag{2.81}$$

for a $Z - Z$ electrolyte, with

$$\frac{\partial \Psi_1}{\partial z} = 0 \tag{2.82}$$

at the midplane where $Z = 0$. On the surface of the sphere, $Z = g(r)$, the condition is

$$\Psi_1 = \Psi_s \quad or \quad \frac{\partial \Psi_1}{\partial z} = q, \tag{2.83}$$

where

$$g(r) = \frac{1}{2}\kappa h + \frac{1}{2}r^2, \tag{2.84}$$

and the charge is scaled on $\varepsilon \varepsilon_0 kT\kappa/ez$.

The problem has been reduced to one dimension but further analytical progress requires linearization of the differential equation, *i.e.* small potentials. The force derived from equation 2.77,

$$F \approx \pi \varepsilon \varepsilon_0 \left(\frac{kT}{ze}\right)^2 \int_0^\infty \left[(\kappa a)\Psi_1^2 + \left(\frac{\partial \Psi_1}{\partial r}\right)^2\right] r\, dr, \tag{2.85}$$

$$Constant\ charge: F \approx 2\pi\varepsilon\varepsilon_0 \left(\frac{kT}{ze}\right)^2 \kappa a\, q^2 \frac{\exp(-\kappa h)}{1 - \exp(-\kappa h)} \qquad (2.86)$$

Extensions of the analytical solution show that the error for constant potential boundary conditions remains small as the gap is diminished. Conversely, the terms neglected in the constant charge calculation grow without bound, showing that this approximation is invalid when the gap is much smaller than the Debye thickness. This problem stems from the radial gradients in the potential neglected in equation 2.80.

Linear superposition of single sphere potentials also provides a useful approximation for the repulsive force. From the Debye-Hckel solution around a single sphere, equation 2.62, we find

$$F \approx \pi\varepsilon\varepsilon_0 \left(\frac{kT}{ze}\right)^2 \Psi_s^2 \frac{1 + \kappa(h + 2a)}{(h/2a + 1)^2} \exp(-\kappa h) \qquad (2.87)$$

for the force. For spheres with thin double-layers, equation 2.59 yields

$$F \approx 32\pi\varepsilon\varepsilon_0 \left(\frac{kT}{ze}\right)^2 \kappa a \tanh^2(\tfrac{1}{4}\Psi_s) \exp(-\kappa h) \qquad (2.88)$$

By solving the linearized Poisson-Boltzmann equation through a multipole expansion, Russel *et al.* mapped out regions where the Derjaguin and linear superposition approximations are valid with small potentials.

For thin double-layers, the situation for which it is intended, the Derjaguin approximation produces very accurate results for constant-potential boundary conditions when the potential is derived from the one-dimensional non-linear Poisson-Bolzmann equation.

Attention has been centered on the repulsive force due to electrostatic interactions since this is measured directly in many experiments. However, it is also necessary to know the electrostatic interaction energy, Φ, defined as

$$F = -\frac{\partial\Phi}{\partial h}. \qquad (2.89)$$

Interaction energies

Geometry	Constraint	Force expression	Φ
Two flat plates	Superposition	(2.66)	$64kTn_b\kappa^{-1}\tanh^2(\tfrac{1}{4}\Psi_s)\exp(-\kappa h)$

Two spheres	Constant potential	(2.85)	$2\pi\varepsilon\varepsilon_0 \left(\frac{kT}{ze}\right)^2 a\Psi_s^2 \ln(1 + e^{-\kappa h})$
Two spheres	Constant charge	(2.86)	$-2\pi\varepsilon\varepsilon_0 \left(\frac{kT}{ze}\right)^2 aq^2 \ln(1 - e^{-\kappa h})$
Two spheres	Linear superposition	(2.87)	$4\pi\varepsilon\varepsilon_0 \left(\frac{kT}{ze}\right)^2 \frac{a^2}{h+2a} \Psi_s^2 \exp(-\kappa h)$
Two spheres	Superposition	(2.88)	$32\pi\varepsilon\varepsilon_0 \left(\frac{kT}{ze}\right)^2 a \tanh^2(\frac{1}{4}\Psi_s) \exp(-\kappa h)$

This energy can be calculated from repulsive force by integration. Interaction energies (or potentials) are used extensively when dealing with colloid stabilly. The Table lists approximate forms.

Bibliography

[1] R.P. Feynman, R.B. Leighton and M. Sands *The Feynman Lectures on Physics* vol. II, Addison-Wesley, New York, NY, 1964.

[2] L.D. Landau and E.M. Lifschitz *Electrodynamics of continuous media* Pergamon, Oxford, 1960.

[3] D.A. MacQuarrie *Statistical Mechanics* Harper and Row, New York, N.Y., 1976.

[4] G. Gouy *J. Phys. Radium* 9 (1910) 457.

[5] D.L. Chapman *Phil. Mag.* 25 (1913) 475.

[6] D.A. Haydon The electrical double-layer and electrokinetic phenomena, in *Progress in Surface Science*, vol. I. Academic press., London, 1964.

[7] M.J. Sparnaay *The electrical double layer*, Pergamon, Oxford, 1972.

[8] W.B. Russel, D.A. Saville and W.R. Schowalter *Colloidal dispersions* Cambridge, 1989.

[9] R.J. Hunter *Introduction to modern colloid science*, Oxford, 1993.

[10] A.L Loeb, J.T.G. Overbeek and P.H. Wiersema *The Electrical Double-Layer around a Spherical Colloid particle*, MIT Press Cambridge Mas., 1961.

[11] B.P. Lee and M.E. Fisher *Phys. Rev. Let.* 76 (1996) 2906.

[12] B.P. Lee and M.E. Fisher *Europhys. Let.* 39 (1997) 611.

[13] D.M. Zuckerman, M.E. Fisher and B.P. Lee *Phys. Rev. E* 56 (1997) 6569.

[14] M.E. Fisher and D.M. Zuckerman *J. Chem. Phys.* 109 (1998) 7961.

[15] M.E. Fisher and D.M. Zuckerman *Chem. Phys. Let.* 293 (1998) 461.

[16] Y. Levin and M.E. Fisher *Physica A* 225 (1996) 164.

[17] Xiaojun Li, Y. Levin and M.E. Fisher *Europhys. Let.* 26 (1994) 683.

[18] M.E. Fisher, Y. Levin and Xiaojun Li *J. chem. Phys.* 101 (1994) 2273.

[19] Y. Levin and M.C. Barbosa *J. Phys. II* 7 (1997) 37.

[20] M. N. Tamashiro, Y. Levin and M.C. Barbosa *Physica A* 258 (1998) 341.

[21] J.M.G. Barthel, H. Krienke and W. Kunz *Physical chemistry of electrolyte solutions*, Topics in Physical Chemistry 5, Springer, Darmstadt, 1998.

Chapter 3

Van der Waals forces

3.1 Introduction

Microscopic observations of colloidal particles in the nineteenth century have shown their tendency to form persistent aggregates, even for uncharged particles without specific reactive site. This behaviour indicates attractive interparticle force. This attraction arises from dipolar interaction: local fluctuations in the polarization within one particle induces correlated response in the others via the propagation of electromagnetic waves. A general description, with many-body interaction, is very complicated. Thus, de Boer [1] and Hamaker [2] assumed the intermolecular forces to be strictly pairwise additive. Later, Lifshitz et al [3] proposed a continuum theory where many-body effects are taken into account by treating the particles and the subphase as individual macroscopic phases characterized by their dielectric properties. The basis of this treatment lies in quantum electrodynamic theory but an easier description was reformulated by van Kampen et al [4] and the first quantitative implementation of the theory to real systems began with Parsegian and Ninham [5]. In vacuum, the solution of Laplace's equation for the electric potential due to a point dipole $\mathbf{p}_1$ is given by

$$\phi_1\left(\mathbf{r}\right) = \frac{1}{4\pi\epsilon_o}\mathbf{p}_1\nabla\left(\frac{1}{r}\right) \qquad (3.1)$$

If a second dipole of moment $\mathbf{p}_2$ is placed in this field at position $\mathbf{r}$, an interaction energy U results which reads

$$U = \mathbf{p}_2.\nabla\phi_1 \qquad (3.2)$$

Molecules that do not possess permanent dipole posses a non-zero instantaneous dipole moment because of fluctuations caused for instance by electromagnetic radiation fields. Then instantaneous values of dipolar moment must be taken into

47

account: $\mathbf{p}_{inst}$. This dipole can then induce a dipolar moment $\mathbf{p}_{ind}$ in a second molecule which is a function of its polarisability α. Also, this kind of interaction exists even for unpolar molecules.

Let us go back to the definition of the electric dipolar moment $\mathbf{p}_{inst}$ in a molecule. The instantaneous configuration of a molecule consisting of a set of nuclei and electrons with charges q_i at positions $\mathbf{r}_i$. The net charge is given by

$$Q = \sum_i q_i \tag{3.3}$$

and the electric dipole moment is defined as

$$\mathbf{p} = \sum_i q_i \mathbf{r}_i. \tag{3.4}$$

When two molecules A and B are brought from an infinite separation to a distance R, the charges on each particle will interact. The interaction energy $V_{int}(R)$ is defined by

$$V_{int}(R) = E^{AB}(R) - E_0^A - E_0^B \tag{3.5}$$

with E_0^A and E_0^B the ground state energy of each particle and $E^{AB}(R)$ the energy of the new system. The interaction energy operator, $\mathcal{V}_{int}$ in the total Hamiltonian of the system can be written as

$$\mathcal{V}_{int} = \frac{1}{4\pi\varepsilon_0} \sum_i \sum_j \frac{q_i^A q_i^B}{\left|\mathbf{R} - \mathbf{r}_j^B - \mathbf{r}_i^A\right|} \tag{3.6}$$

in a vacuum, with $\mathbf{R}$ the position of the center of mass of the molecule B relative to the center of mass of the molecule A. We suppose here that there is no spin-spin interaction and that the electromagnetic interaction is an electrostatic one.

If the intermolecular distance R is greater than $\left|\mathbf{r}_i^A\right|$ and $\left|\mathbf{r}_j^B\right|$, the last equation can be expanded in powers of $1/R$.

For neutral particles,

$$\sum_i q_i = 0 \quad \sum_j q_j = 0 \tag{3.7}$$

and the terms with only $\mathbf{r}_i$ or $\mathbf{r}_j$ are equals to zero and

$$\mathcal{V}_{int} = \frac{1}{4\pi\varepsilon_0} \left(\frac{\mathbf{p}^A \cdot \mathbf{p}^B}{R^3} - 3\frac{\left(\mathbf{p}^A \cdot \mathbf{R}\right)\left(\mathbf{p}^B \cdot \mathbf{R}\right)}{R^5} + \mathcal{O}\left(\frac{1}{R^4}\right) \right) \tag{3.8}$$

If A and B are polar molecules,

$$\langle \mathbf{p} \rangle_0 = \mu \neq 0 \tag{3.9}$$

with μ the permanent dielectric moment.

If A and B are non-polar molecules, it is necessary to calculate $E^{AB}(R)$ to the second order i.e. to consider how V_{int} causes the internal state of each molecule to change.

At this approximation order,

$$V_{int}^{(2)}(R) = -\sum_{n,m,\neq 0} \frac{|\langle 0,0|V_{int}|n,m\rangle|^2}{E_m^A + E_n^B - E_0^A - E_0^B} \tag{3.10}$$

where

$$|m,n\rangle = |m^A\rangle |n^B\rangle \tag{3.11}$$

If A and B are symmetrical, we can rewrite eq. 3.10 as

$$V_{int}^{(2)}(R) = -\frac{C_{AB}}{R^6} \tag{3.12}$$

where

$$C_{AB} = \frac{3e^4\hbar}{2m_e^2(4\pi\varepsilon_0)^2} \sum_{n,m,\neq 0} \frac{f_{0m}^A f_{0n}^B}{\omega_{0m}^A \omega_{0n}^B (\omega_{0m}^A + \omega_{0n}^B)} \tag{3.13}$$

with

$$\hbar = \frac{h}{2\pi} \qquad h = 6.626 \, 10^{-34} \, \text{Js}^{-1} \tag{3.14}$$

and

$$\omega_{0m} = (E_m - E_0)/\hbar \tag{3.15}$$

is the pulsation of the electromagnetic radiation that would cause the transition from ground state $|0\rangle$ to the excited state $|m\rangle$ of the isolated molecule.

$$f_{0m} = \frac{2m_e\omega_{0m}}{\hbar e^2} |\langle 0|p_z|m\rangle|^2 \tag{3.16}$$

p_z is the z co-ordinate of the dipole moment operator with a spherical assumption $\langle 0|p_z|m\rangle = \langle 0|p_y|m\rangle = \langle 0|p_z|m\rangle$. This expression of $V_{int}(R)$ is the first term of the infinite series

$$V_{int}(R) = -\sum_{l_1 \geq 1}\sum_{l_2 \geq 1} \frac{C_{l_1 l_2}}{R^{2(l_1+l_2+1)}} \tag{3.17}$$

Concerning colloidal science, it has been shown that the term $l_1 = l_2 = 1$ is the leading term [6]. A description of the unsymmetric case has been given in the same reference.

The next picture gives the main interaction formulas for charged and uncharged particles.

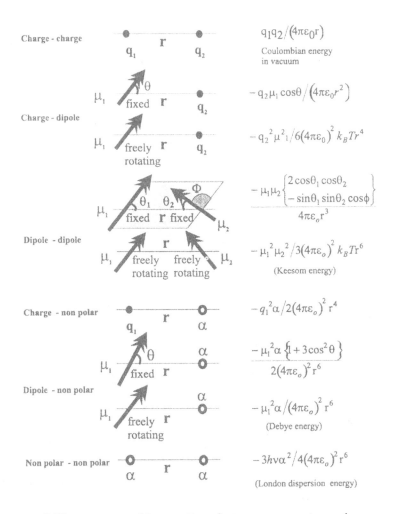

Figure 3.1: Different types of interactions between some atoms, ions or molecules in the vacuum. q is the electric charge, μ is the electric dipole moment, α is the electric polarisability, r is the distance between interacting particles, ν (s^{-1}) is the electronic absorption frequency i.e. if I is the ionization potential, $I = h\nu$, T is the absolute temperature, k_B is the Boltzmann constant ($k_B \equiv 1.38\,10^{-23}\ JK^{-1}$) and ϵ_0 is the absolute permitivity ($\epsilon_0 \equiv 8.854\,10^{-12}\ C^2J^{-1}m^{-1}$).

3.2 Interaction between polar molecules (small particles)

We can rewrite the interaction free energy as :

$$V(r) = -C_{12}/r^6 = -\{C_{ind} + C_{orient} + C_{disp}\}/r^6 \qquad (3.18)$$

where the coefficients due to the induction (C_{ind}), the orientation (C_{orient}) and the dispersion (C_{disp}) are given by:

$$C_{ind} = \left(\mu_1^2\alpha_{02} + \mu_2^2\alpha_{01}\right)/(4\pi\epsilon_0)^2 \qquad (3.19)$$

$$C_{orient} = \left(\mu_1^2\mu_2^2\right)/\left(3k_BT(4\pi\epsilon_0)^2\right) \qquad (3.20)$$

and

$$C_{disp} = \left(3h\nu_1\nu_2\alpha_{01}\alpha_{02}\right)/\left(2(\nu_1 + \nu_2)(4\pi\epsilon_0)^2\right) \qquad (3.21)$$

In most cases, the dispersion forces are dominant except for small and highly polar molecules like water. These expressions are in good agreement with experimental data and when the studied interaction arises between dissimilar molecules 1 and 2, C_{12} is close to the geometric mean of C_{11} and C_{22}.

The important case of water does not obey this empiric law because it is an highly polarised molecule.

The London theory assumes that molecules have only one single ionization potential and therefore one absorption frequency and it does not handle the effect of the solvent to the particle interaction potential.

In 1963, McLachlan [7, 8] gave a new expression for the Van der Waals free energy of two small particles (1 and 2) in a medium (3):

$$V(r) = -\frac{3k_BT}{(4\pi\epsilon_0)^2\,r^6}\frac{\alpha_1(0)\alpha_2(0)}{\epsilon_3^2(0)} - \frac{6k_BT}{(4\pi\epsilon_0)^2\,r^6}\sum_{n=1}^{\infty}\frac{\alpha_1(\imath\nu_n)\alpha_2(\imath\nu_n)}{\epsilon_3^2(\imath\nu_n)} \qquad (3.22)$$

where $\alpha_j(\imath\nu_n)$ is the polarisability of the molecule j at the imaginary frequency $\imath\nu_n$ and $\epsilon_3^2(\imath\nu_n)$ is the relative dielectric permittivity of the medium at the same frequency and

$$\nu_n = n\,(k_BT/\hbar) \qquad (3.23)$$

All these quantities can be measured independently.

If we consider only the zero frequency and a particle j with a permanent dipole moment μ_j, $\alpha_j(\imath\nu_n)$ is reduced to the Debye-Langevin equation:

$$\alpha_j(0) = \frac{\mu_j}{3k_BT} + \alpha_{0j} \qquad (3.24)$$

In vacuum equation 3.22 becomes:

$$
\begin{aligned}
V(r) &= -\frac{3k_BT}{(4\pi\epsilon_0)^2\, r^6}\alpha_1(0)\alpha_2(0) \\
&= -\frac{3k_BT}{(4\pi\epsilon_0)^2\, r^6} \left(\frac{\mu_1}{3k_BT} + \alpha_{01}\right)\left(\frac{\mu_2}{3k_BT} + \alpha_{02}\right) \qquad (3.25)
\end{aligned}
$$

And we find again the equation 3.18.

Van der Waals forces have peculiar properties. As the polarisability is generally anisotropic (except for ideal spherical particles), Van der Waals forces are anisotropic too. However, the orienting effects of the anisotropic dispersion forces are usually less important than other forces like dipole-dipole interactions. Another problem in the description of these forces for a system is that Van der Waals forces are not generally pairwise additive. This property is very important for large particles and surfaces in a medium. The last specificity is the retardation effect. Concerning the dispersion energy: as the speed of the interaction between particles is limited by the speed of light ($\equiv 3\,10^8\,\mathrm{ms}^{-1}$ in a vacuum), the states of the fluctuating dipoles are not the same when the fisrt one sends the information as when the second one receives it. The consequence is that the power law could be closer to $-1/r^7$ than $-1/r^6$. This latter effect is more important in a medium where the speed of light is slower than in the vacuum and can become very important for macroscopic bodies. When the size of the body becomes greater than a distance where this decay induces a change in the power law, retardation effect must be taken into account.

For a more detailed description on this topic the interested reader can be referred to refs [9, 10]

3.3   Interaction between surfaces (big particles)

The three most important forces for the long range interaction between macroscopic particles and a surface are steric-polymer forces, electrostatic interactions and Van der Waals forces. If we assume than the Van der Waals interactions between two atoms in a vaccuum are non-retarded and additive, we saw in the previous chapter that the form of the Van der Waals pair potential is: $w = -C/D^6$ where C is the coefficient in the atom-atom pair potential and D is the distance between the two atoms. We can then integrate the energy of all the atoms wich form the surface and the studied atom. In the same way, we can calculate interactions between surfaces with different geometries making the integration for all the atoms on each surface. The resulting interaction law is given for different geometries in the next figures where the Hamaker constant is introduced:

$$A = \pi^2 C \rho_1 \rho_2 \qquad (3.26)$$

where ρ_i is the atom density (number of atom per unit volum) of each body.

Figure 3.2: Non-retarded van der Waals interaction free energy w between two atoms. $w = -C/D^6$ where C is the coefficient in the atom-atom pair potential (eq. 3.18).

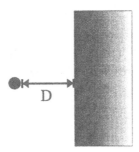

Figure 3.3: Non-retarded van der Waals interaction free energy w between an atom and a flat surface. $w = -\pi C \rho / (6D^3)$ where ρ is the number of atoms per unit volume, C is given by eq. 3.18 and D is the distance between the atom and the surface.

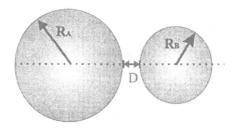

Figure 3.4: Non-retarded van der Waals interaction free energy w between two spheres. $w = -A/(6D)(R_1R_2/(R_1 + R_2))$ where R_1 and R_2 are their radii and D is the distance between their surfaces. $A = \pi^2 C \rho_1 \rho_2$ is the Hamaker constant with ρ_1 and ρ_2 the number of atoms per unit volume of each sphere and C is given by eq. 3.18.

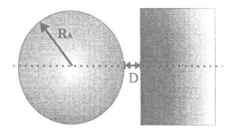

Figure 3.5: Non-retarded van der Waals interaction free energy w between a sphere and a flat surface. $w = -(A/6D)R$ with R the radius of the sphere, D the distance between the two surfaces and A the Hamaker constant (fig. 3.4).

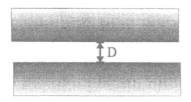

Figure 3.6: Non-retarded van der Waals interaction free energy w per unit area between two flat surfaces. $w = -A/(12\pi D^2)$ with D the distance between the two surfaces and A the Hamaker constant (fig. 3.4)

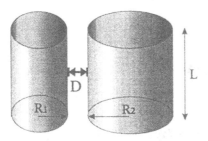

Figure 3.7: Non-retarded van der Waals interaction free energy between two parallel cylinders. $w = -\left\{AL/\left(12\,2^{1/2}\,D^{3/2}\right)\right\}\left(R_1R_2/\left(R_1+R_2\right)\right)^{1/2}$ with R_1 and R_2 the radii of each cylinder, L their length, D the distance between the two surfaces and A the Hamaker constant (fig. 3.4).

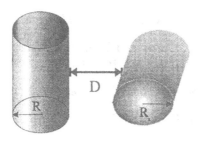

Figure 3.8: Non-retarded van der Waals interaction free energy between two perpendicular cylinders. $w = -A/\left(6D\right)\left(R_1R_2\right)^{1/2}$ with D the distance between the two surfaces, R_1 and R_2 the radii of each cylinder and A the Hamaker constant (fig. 3.4).

Bibliography

[1] J.H. de Boer *Trans. Faraday Soc.* 32 (1936) 10.

[2] H.C. Hamaker *Physica* 4 (1937) 1058.

[3] E.M. Lifshitz *et al. Soviet Physics JETP* 2 (1956) 73.; I.E. Dzaloshinskii, E.M. Lifsitz and L.P. Pitaevskii *Adv. Phys.* 10 (1961) 165.

[4] N.G. van Kampen, D.R.A. Nijboer and K. Schram *Phys. Lett.* 26A (1968) 307.

[5] V.A. Parsegian and B.W. Ninham *Nature (London)* (1969) 1197.

[6] J.O. Hirschfelder, C.F. Curtis and R.B. Bird *Molecular theory of gases and liquids* Wiley N.Y., Chapman and Hall, London, 1954.

[7] A. D. McLachlan *Proc. Roy. Soc. Lond.* Ser. A202 (1963) 224.

[8] A. D. McLachlan *Mol. Phys.* 6 (1963) 423.

[9] J. Israelachvili *Intermolecular and surface forces*, Academic Press, San Diego, 1992.

[10] W.B. Russel, D.A. Saville and W.R. Schowalter *Collidal dispersions*, G.K. Batchelor, Cambridge University Press, Cambridge, 1989.

Part II

Bulk Electrolytes

Chapter 1

Introduction

1.1 Introduction

We study electrolytic solutions in which the ions are represented by charged hard spheres and the solvent (water in most cases) by neutral hard spheres with a charge distribution or a continuum. We are mainly interested in models that admit analytical solutions in some approximations. This is certainly the case with the mean spherical approximation (MSA), which is connected to the Debye Hückel limiting law (DHLL) since both are solutions of the linearized Poisson Boltzmann equation. The main simplifying assumption of the DHLL is that the ions in the neutralizing ionic cloud around each ion are taken as point ions. The MSA is the solution of the same linearized Poisson Boltzmann equation but with finite size ions in the cloud. The mathematical solution of the proper boundary conditions of this problem is much more complex. However, simple variational derivations exist nowadays [1, 2, 24].

The analytic solution of the MSA shares with the DHLL the remarkable simplicity of being a function of a single screening parameter Γ [3] for an arbitrary neutral mixture of ions, and also for a variety of non spherical associated charged [4, 5] objects. The expressions of the thermodynamic excess functions are formally very similar to those of the DHLL. The MSA can be derived from first principles of statistical mechanics. Better approximations are the HNC equation and its improvements

59

but they need to be solved numerically for every individual system, which is often impractical. The MSA has been solved for the so-called 'primitive' model [6], in which the solvent is regarded as a dielectric continuum, and for the 'non-primitive' model [7]-[12] in which the solvent is discrete and modeled as a collection of hard spheres with embedded point dipoles.

Calculations of departures from ideality in ionic solutions using the MSA have been published in the past by a number of authors. Effective ionic radii have been determined for the calculation of osmotic coefficients for concentrated salts [13], in solutions up to 1 mol/L [14] and for the computation of activity coefficients in ionic mixtures [15]. In these studies, for a given salt, a unique hard sphere diameter was determined for the whole concentration range. Also, thermodynamic data were fitted with the use of one linearly density-dependent parameter (a hard core size $\sigma(C)$, or dielectric parameter $\varepsilon(C)$), up to 2 mol/L, by least-squares refinement [16]-[18], or quite recently with a non-linearly varying cation size [19] in very concentrated electrolytes.

Liquid state theories like the MSA (mean spherical approximation) and the HNC (hypernetted chain approximation) can be derived as variational problems [1, 2] for the free energy functional, which is written in terms of a single screening parameter Γ and which interpolates between the low coupling Debye-Hückel and high coupling, high density [1, 20, 21], and low density [22, 23, 24] limits for systems of hard objects in general, and hard spheres in particular, and which can be mapped onto the solution of the OZ (Ornstein-Zernike) equation of the MSA (mean spherical approximation) [25, 26, 27, 3]. The idea, furthermore is that in the MSA, both the thermodynamics and the structure can be represented by a simple geometrical model consisting of a capacitor, generally spherical, but also of any given shape.

We are interested in simple yet realistic models of ionic solutions. These models admit analytic solutions, and in general with simple expressions. They can be classified in three categories:

1. The primitive model, in which the ions are charged spheres, and the solvent is a dielectric continuum;

2. the elementary model, in which the solvent is a point dipole in a hard sphere;

3. the basic model, in which we add a potential of tetrahedral octupolar symmetry to mimic the hydrogen bonds.

We will address only the first category here. However, the parametrization of experimental data will keep in mind the consequences of (2) and (3), namely that the dielectric screening will depend on all solutes. The current status is that (1) is well understood, (2) is being developed and (3) is just barely being being started, but we have good indications that these models give good agreement with structural data.

Ionic solutions are liquids consisting of a solvent formed from neutral, polar molecules, and a solute that dissociates into positive and negative ions. They vary widely in complexity: in the classic electrolyte solutions, the cations and anions are of comparable size and absolute charge, whereas macromolecular ionic solutions contain both macroins (charged polymer chains or coils, micelles, charged colloidal particles etc.) and microscopic counterions [28]. We will discuss only the classic ionic solutions.

In the asymptotic limit of strong Coulomb interactions between the charged particles , that is the limit in which either the charge goes to infinity or the temperature goes to zero, is the starting point of our present discussion [29]: The free energy and the internal energy diverge to the same order in the coupling parameter while the entropy diverges at a slower rate. In this asymptotic limit, the free energy and the energy coincide, and furthermore, the mean spherical approximation (MSA) and the hypernetted chain approximation (HNC) coincide. In the asymptotic limit the excess electrostatic energy is identical to the exact Onsager lower bound, which is achieved by immersing the entire hard core system in an infinite neutral and perfectly conducting (liquid metal) fluid. The Onsager process of introducing the infinite conductor, naturally decouples all the components in the system which may differ in size, shape, charge distribution and relative orientation in space. As a result, the variational free energy functional in the high coupling limit diagonalizes, and the mathematical solution of the asymptotic problem is given in terms of geometrical properties of the individual particles in the system.

As an illustration, consider the charges induced on the surface of each particle when placed in an infinite conductor. Then the direct correlation function in the asymptotic strong coupling limit (Onsager picture) is obtained directly from the electrostatic interaction of the charges of the particles smeared on the surface of those particles. The calculation of the bridge function, (the part missing in the HNC approximation) involves the construct of Onsager molecules for the potential of mean force. Another asymptotic limit is the high density limit, in which the compressibility tends to zero because of the tight packing of the particles. In this case the MSA solution is also obtained from a simple geometric argument by computing the overlap volume of the particles as a function of their distance and their relative orientation. These two distinct limits provide the set of basis functions for the representation of the direct correlation function, which can be shown to be sufficient to represent the dcf of the complete MSA solution. In other words, these two limits provide the full functional basis set for the exact solution o an asymptotic approximation of the HNC solution for all densities and temperatures, for hard charged objects. This enables us to replace a functional variational problem by a variational problem in which the basis functional set is fixed, and known, and where we only need to find the weights of these basis functions. The asymptotic Onsager state of the system is essentially the analog of the diagonalizable reference Hamiltonian of quantum mechanics, when the Schroedinger equation has to be solved. In our case the basis functions for the functional expansion of the direct correlation function are obtained from linear combinations of overlap functions, such as the volume, the surface and the convex radius, and the electrostatic interaction between surface s-meared charges. The full solution is obtained by associating free parameters with various parts of the basis functions. By proper manipulation of the free parameters, and by a judicious selection of the basis set of trial functions, one can obtain, as in quantum mechanics, different levels of approximations. The physically intuitive meaning of the basis functions in the representation of the dcf is particularly illuminating in the formulation of perturbation treatments. The use of the asymptotic basis set of functions ensures that at all levels of the perturbation approximation, the resulting free energy has the desired property of interpolating between two exact lower bounds, the DH result (which is effective at weak coupling) and the Onsager result, (which is effective at high coupling). These two limits pin the free energy. The seemingly complex direct correlation function for the MSA of ionic mixtures obtained by Hiroike [30] can be written in terms of the basis functions mentioned above. We then provide more insight into the nature and physical meaning of the solution as represented in the resulting free energy. Finally we point out possible directions along which free energy models of the inhomogenous fluid can be constructed.

The only condition imposed on the system is the electroneutrality condition:

$$\sum_i^m \rho_i z_i = 0 \tag{1.1}$$

where ρ_i is the number density of spheres of type i with diameter σ_i and charge z_i in electron units, and m the total number of species.

1.1.1 The primitive model and Debye-Hückel (DH) theory.

Consider a neutral mixture of charged hard spheres of diameter σ_i, charge $z_i e$ (where e is the charge of the electron), number density $\rho_i = \frac{N_i}{V}$, (where N_i is the number of ions of species i enclosed in a volume V). The number of species is s. where ρ_i is expressed in number of particles per cubic angstrom and ν_i is the stoichiometric coefficient. We will use Boltzmann's constant k_B and write $\beta = \frac{1}{k_B T}$.

The charge distribution around ion i is

$$q_i(r) = e \sum_{j=1}^{s} z_j \rho_j^{(i)}(r) \tag{1.2}$$

where $\rho_j^{(i)}(r)$ is the conditional density of ions j in the neighborhood of i. In statistical mechanics this function is usually expressed in terms of the pair correlation function

$$g_{ij}(r) = \frac{\rho_j^{(i)}(r)}{\rho_j} = g_{ji}(r) = \frac{\rho_i^{(j)}(r)}{\rho_i} \tag{1.3}$$

which is a symmetric function in the exchange of particles i and j From a simple analogy to the atmospheric pressure equation, the density $\rho(r)$ of the atmosphere is given by

$$\rho(r) = \rho_0 e^{-\beta m g r} \tag{1.4}$$

where the term mgr represents the potential energy of a particle of mass m at a height r, g is the acceleration of gravity.

More generally $\rho(r)$ becomes $\rho_j^{(i)}(r)$, particle i is the earth.

We define the potential of mean force $w_{ij}(r)$ such that

$$g_{ij}(r) = e^{-\beta w_{ij}(r)} \tag{1.5}$$

Our central problem will be to determine this potential of mean force . There are a number of requirements on it, however.

We write:

$$w_{ij}(r) = e z_j \varphi_i(r) + \zeta_{ij}(r) \tag{1.6}$$

where the first term is purely electrostatic and $\zeta_{ij}(r)$ contains all the remaining contributions, such as excluded volume. In other words

$$\begin{aligned} g_{ij}(r) &= e^{-\beta e z_j \varphi_i(r)} \\ &\simeq 1 - \beta e z_j \varphi_i(r). \end{aligned} \tag{1.7}$$

If we assume $\zeta_{ij}(r) \cong 0$, then the electrostatic potential must satisfy the Poisson equation in the form (the Poisson equation is valid also when $\zeta_{ij} \neq 0$)

$$\nabla^2 \varphi_i(r) = -\frac{4\pi}{\epsilon_0} q_i(r) \tag{1.8}$$

and using (1.2) and (1.5)

$$\nabla^2 \varphi_i(r) = -\frac{4\pi}{\epsilon_0} e \sum_{j=1}^{s} \rho_j z_j e^{-\beta e z_j \varphi_i(r)} \tag{1.9}$$

which is the Poisson-Boltzmann or Milner equation [31]. Even for the simplest possible case the equal size and equal charge electrolyte ($\sigma_+ = \sigma_- = \sigma$, $z_+ = -z_-$) this equation cannot be solved in closed form. But there are asymptotic regimes in which we can solve it. If σ is very small the conditional probability density must be of the form:

$$\lim_{\sigma \to 0} \rho(\sigma) = e^{-\beta e z_j \varphi_i^0(r)} \tag{1.10}$$

where $\varphi_j^0(r) = \frac{e z_i}{\epsilon_0 r}$ is the bare Coulomb interaction, or in other words, when two charges come very close, their own interaction will dominate over the interactions of the other surrounding charges.

Another limiting case is when the central ion is very large. Then $\varphi_i(r)$ must be small, and we know that, for $r \to \infty$ then:

$$g_{ij}(r) = 1$$
$$w_{ij} \to 0$$
$$\varphi_{ij}(r) \to 0 \tag{1.11}$$

and, following Debye and Hückel (DH) we can expand the exponential in Eq. (1.9), the result being [32]:

$$\nabla^2 \varphi_i(r) = \kappa^2 \varphi_i(r) \tag{1.12}$$

where

$$\kappa^2 = \frac{4\pi \beta e^2}{\epsilon_0} \sum \rho_j z_j^2 \tag{1.13}$$

which defines the Debye screening length. There are several ways of solving the Eq. (1.12). Consider, for simplicity only the restricted case in which all ions are of equal size. Then, the distance of closest approach is σ. We have to transform the gradient to spherical coordinates, but since, φ does not depend on the angles, we simply get:

$$\frac{1}{r} \frac{\partial^2}{\partial r^2} r \varphi_i(r) = \kappa^2 \varphi_i(r) \tag{1.14}$$

The general solution of this equation is:

$$r\varphi_i(r) = A_i e^{-\kappa(r-\sigma)} + B_i e^{\kappa(r-\sigma)}. \tag{1.15}$$

The only way to satisfy boundary condition (1.11) is to require that $B_i = 0$. The value of A_i is obtained form Gauss's theorem, or more simply, from boundary condition (1.10), since for $r \to 0$

$$r\varphi_i(r) = \frac{z_i e}{\epsilon_0} \left(\frac{1}{1+\kappa\sigma} \right) \tag{1.16}$$

The full solution for the potential is

$$\varphi_i(r) = \frac{z_i e}{\epsilon_0} \frac{e^{-\kappa(r-\sigma)}}{r(1+\kappa\sigma)} \tag{1.17}$$

and according to (1.8) and (1.14) the charge density is:

$$
\begin{aligned}
q_i(r) &= -\frac{\epsilon_0}{4\pi}\nabla^2\varphi_i(r) = -\frac{z_i e}{4\pi r}\frac{\partial^2}{\partial r^2}e^{-\kappa(r-\sigma)} \\
&= -\frac{z_i e}{4\pi}\frac{\kappa^2}{r}e^{-\kappa(r-\sigma)}.
\end{aligned} \tag{1.18}
$$

This charge distribution satisfies the electroneutrality condition

$$-z_i e = \int d\mathbf{r}\, q_i(\mathbf{r}) \tag{1.19}$$

which can be verified by integration: substitution of (1.18) into (1.19) leads to

$$\sum_j z_j 4\pi \int_0^\infty dr r^2 h_{ij}(r)\rho_j = -z_i \tag{1.20}$$

This is a completely general, and rigorous sum rule[33] , that must be obeyed by the distribution functions of any good theory of electrolytic solutions. It means that the charge of the ionic cloud surrounding a given ion just has enough charge to neutralize that ion. The charge distribution, in our theory, is exponentially decaying with a mean distance of decay equal to $\frac{1}{\kappa}$.

The excess energy of charging up the system is from (1.18) and Coulomb's formula (1.10 ff)

$$
\begin{aligned}
\Delta E^{ch} &= \frac{1}{2}\left[\frac{4\pi e^2}{\epsilon_0}\right]\sum_{i,j}\rho_i\rho_j z_i z_j \int dr\, r^2 g_{ij}(r)\frac{1}{r} \\
&= -\sum_i \frac{\rho_i(e z_i)^2}{2\epsilon_0}\frac{1}{\sigma+\frac{1}{\kappa}}
\end{aligned} \tag{1.21}
$$

which is the energy of a system of a spherical capacitors of radius $(\sigma + \frac{1}{\kappa})$. The internal energy has a lower bound, due to Onsager [34]; imagine that we increase κ (eq. (1.13)), either by charging up the ions ($z_i \to \infty$) or by letting the temperature drop ($\beta \to \infty$) and letting the density $\rho_i \to \infty$. Physically, this is equivalent to inmersing all our ions in liquid metal. Then the screening length $\frac{1}{\kappa}$ is zero: the system is a perfect screening system, and the energy

$$\Delta E^{ch} = -\sum_i \frac{\rho_i (e z_i)^2}{2\epsilon_0 \sigma} \tag{1.22}$$

is a rigorous lower bound for the energy of any system of hard charged ions. From

$$\Delta A = \Delta E - T \Delta S \tag{1.23}$$

$$\frac{\partial \Delta A}{\partial T} = -\Delta S \tag{1.24}$$

$$\frac{\partial \Delta A / T}{\partial (1/T)} = \Delta E \tag{1.25}$$

we get

$$\Delta A = \frac{1}{\beta} \int_0^\beta d\beta' \Delta E(\beta') \tag{1.26}$$

From equation (2.22) in the infinite dilution limit we get the Debye-Hückel excess charging energy

$$\Delta E = -\frac{\kappa^3}{8\pi\beta}$$

substituting into (1.26) leads to

$$\Delta A = -\frac{\kappa^3}{12\pi\beta} = -\frac{\kappa^3}{8\pi\beta} + \frac{\kappa^3}{24\pi\beta} = \Delta E + \frac{\kappa^3}{24\pi\beta}. \tag{1.27}$$

And from (2.24)

$$\Delta S = -\frac{\kappa^3}{24\pi T}.$$

The excess osmotic coefficient, defined by:

$$\phi = 1 + \frac{\beta \Delta P}{\rho} \tag{1.28}$$

with ρ =salt concentration = ρ_i for the restricted case, can be obtained from the free energy, using the relation

$$\left(\frac{\partial \Delta A}{\partial V}\right)_T = -\Delta P. \tag{1.29}$$

After a few operations we get:

$$\phi - 1 = -\frac{\kappa^3}{24\pi \sum_i \rho_i}. \tag{1.30}$$

The excess Gibbs free energy is:

$$\frac{\Delta G}{\sum_i N_i} = \frac{\Delta A}{\sum_i N_i} = \frac{\Delta P}{\sum_i \rho_i} \tag{1.31}$$

and using (1.29) and (1.30) we get for the mean electrostatic activity coefficient $\gamma_\pm$ the following relation:

$$\ln \gamma_\pm = \frac{\Delta G}{\sum_i \rho_i} = \frac{\beta \Delta E}{\sum_i \rho_i} = -\frac{\kappa^3}{8\pi \sum_i \rho_i} \tag{1.32}$$

Bibliography

[1] Y. Rosenfeld and L. Blum *J. Chem. Phys.* 85 (1986) 1556.

[2] E. Velazquez and L. Blum, J. Barthel issue of *J. Mol. Fluids.*(In press)

[3] L. Blum *Mol. Phys.* 30 (1975) 1529

[4] L. Blum, Yu.V. Kalyuzhnyi, O. Bernard, and J.N. Herrera *J. Phys. Cond. Matter* 8 (1996) A143.

[5] O. Bernard and L. Blum *J. Chem. Phys.* 104 (1996) 4746.

[6] L. Blum. in *Theoretical Chemistry, Advances and Perspectives*, H. Eyring and D. Henderson eds., vol. 5; Academic Press, New York, 1980.

[7] L. Blum *J. Stat. Phys.* 18 (1978) 451.

[8] F. Vericat and L. Blum *J. Stat. Phys.*22 (1980) 593.

[9] L. Blum and D.Q. Wei *J. Chem. Phys.* 87 (1987) 555.

[10] L. Blum, F. Vericat and W.R. Fawcett *J. Chem. Phys.* 96 (1992) 3039.

[11] L. Blum and W.R. Fawcett *J. Phys. Chem.* 96 (1992) 408.

[12] L. Blum and W.R. Fawcett *J. Phys. Chem.* 97 (1992) 7185.

[13] S. Watanasiri, M.R. Brulé and L.L. Lee *J. Phys. Chem.* 86 (1982) 292.

[14] W. Ebeling and K. Scherwinski *Z. Phys. Chemie* 264 (1983) 1.

[15] H.R. Corti *J. Phys. Chem.* 91 (1987) 686.

[16] R. Triolo, J.R. Grigera and L. Blum *J. Phys. Chem.* 80 (1976) 1858.

[17] R. Triolo, L. Blum and M.A. Floriano *J. Phys. Chem.* 67 (1976) 5956.

[18] R. Triolo, L. Blum, L. and M.A. Floriano *J. Phys. Chem.* 82 (1978) 1368.

[19] T. Sun, J.L. Lénard and A.S. Teja *J. Phys. Chem.* 98 (1994) 6870.

[20] Y. Rosenfeld *Phys. Rev. Lett.* 63 (1989) 980.

[21] Y. Rosenfeld, D. Levesque and J. J. Weiss *J. Chem. Phys.* 92 (1990) 6818.

[22] O. Bernard and L. Blum *Proceedings of the International Conference on Plasmas,* Boston, 1997.

[23] O. Bernard and L. Blum, unpublished.

[24] E. Velazquez and L. Blum, to be published.

[25] J. K. Percus and G. J. Yevick *Phys. Rev.* 118 (1964) 290.

[26] J. L. Lebowitz and J. K. Percus *Phys. Rev.* 144 (1966) 251.

[27] E. Waisman and J. L. Lebowitz *J. Chem. Phys.* 52 (1970) 4307.

[28] E. J. W. Verwey and J. Th. G. Overbeek *Theory of Stability of Lyophobic Colloids,* Elsevier, Amsterdam, 1948.

[29] L. Blum and Y. Rosenfeld *J. Stat. Phys.* 63 (1991) 1177.

[30] K. Hiroike *Mol. Phys.* 30 (1977) 1195.

[31] S. R. Milner *Philos. Mag.* 23 (1912) 521, 25 (1913) 742.

[32] P. Debye and E. Hückel *Phys. Z.* 24 (1923) 185.

[33] L. Blum, Ch. Gruber, J. L. Lebowitz and Ph. A. Martin *Phys. Rev. Lett.* 48 (1982) 1769.

[34] L. Onsager *J. Phys. Chem.* 43 (1939) 189.

Chapter 2

The mean spherical approximation (MSA) for the equal size primitive model

In the previous section we discussed the simplest possible theory, in which the potential of mean force $w_{ij}(r)$, see (1.6), was set equal to the electrostatic potential. This means that we ignore all the other contributions to the ionic interactions, notably the hard-core, which accounts for two very important effects.

a) They prevent the collapse of the system: classical neutral Coulomb (ionic) systems are unstable, because the $(+)$ and the $(-)$, form pairs of unbounded negative energy. This is a rigorous result in statistical mechanics.

b) The excluded volume effect: only one ion can be placed in a given position in space. In the DH theory, the ions of the screening cloud are points, and do not exclude each other. Clearly, the size of the screening cloud of finite size ions must be larger than the DH cloud.

What we want to do now is to include the hard core effects into the calculation of the structure of the ionic cloud. Or, what is equivalent, to charge up a system of hard spheres. This is the basic idea of the mean spherical approximation. A convenient treatment of mixtures of neutral hard spheres is provided by the Percus-Yevick(PY) theory.

71

Consider now the following approximation: take the OZ equation

$$h_{ij}(r) - c_{ij}(r) = \sum_k \rho_k \int dr_1 h_{ik}(r_1) c_{kj}(|\mathbf{r} - \mathbf{r_1}|) \tag{2.1}$$

and use:

i) The hard core condition for separations $r < \sigma$

$$h_{ij}(r) = -1 \tag{2.2}$$

ii) The "Debye-Hückel" (really, MSA) boundary condition for $r \geq \sigma$

$$c_{ij}(r) = -\beta u_{ij}(r) = -\beta \frac{e^2}{\epsilon} \frac{z_i z_j}{r}. \tag{2.3}$$

In the Debye-Hückel limit of zero ionic zips the Eq. (2.1) can be written

$$h_{ij}(r) = -\frac{\beta e^2}{\epsilon_0} \frac{z_i z_j}{r} - \sum_l \rho_l \int dr_1 h_{ik}(r_1) \left[\frac{\beta e^2}{\epsilon_0} \frac{z_l z_j}{|\mathbf{r} - \mathbf{r_1}|} \right] \tag{2.4}$$

using Eq. (1.7) for $h_{ij}(r)$ we get the integral form of the Poisson Boltzmann equation (1.9) (see also 6.1)

$$-\beta z_j e \, \varphi_i(r) = -\frac{\beta z_i z_j e^2}{\epsilon_0 r} - \sum_k \rho_k \int dr_1 \frac{\beta e^2 z_j z_k}{\epsilon_0 |\mathbf{r} - \mathbf{r_1}|} \varphi_i(\mathbf{r_1}) \tag{2.5}$$

The mathematical solution follows the steps outlined for the case of neutral hard spheres. There is, however, one problem in using the Wiener-Hopf factorization [1]: if we take the Fourier transform of Eq. (2.3), we get

$$\int d\mathbf{r} \frac{e^{i\mathbf{k} \cdot \mathbf{r}}}{r} \cong \frac{1}{k^2} \tag{2.6}$$

which has a double pole at the origin, that is, on the real axis. This violates one of the conditions for the factorization. We may, however get around this difficulty by shifting the poles away from the origin. This is done using

$$c_{ij}(r) = -\beta \frac{e^2}{\epsilon_0} z_i z_j \lim_{\mu \to 0} \frac{e^{-\mu |r|}}{|r|}. \tag{2.7}$$

We can check, that, just as in Eqs. (6.30) and (6.31), (see 6.1), the Fourier transform of (2.7) is

$$\tilde{c}_{ij}(k) = -\beta \frac{e^2 z_i z_j}{\epsilon_0} 4\pi \frac{1}{k^2 + \mu^2} |_{\mu \to 0} \tag{2.8}$$

which has two poles located at $k = \pm i\mu$. The Fourier transform of the Ornstein-Zernike equation

$$\lim_{\mu \to 0} \sum_k [\delta_{ik} + \rho_i h_{ik}(k)][\delta_{kj} - \rho_k c_{kj}^\circ + \frac{4\pi\beta e^2}{\epsilon_0} \rho_k \frac{z_k z_j}{k^2 + \mu^2}] = \delta_{ij}. \tag{2.9}$$

This is a matrix equation and, therefore, complicated. If we restrict our analysis to the symmetric 1-1 electrolyte of equal size ions then, we have the symmetries

$$h_{11} = h_{22}, \quad h_{12} = h_{21}$$

$$\rho_1 = \rho_2 = \rho \tag{2.10}$$

In that case the OZ equation can be written as

$$\begin{pmatrix} h_{11} & h_{12} \\ h_{21} & h_{22} \end{pmatrix} - \begin{pmatrix} c_{11} & c_{12} \\ c_{21} & c_{22} \end{pmatrix} = \rho \begin{pmatrix} h_{11} & h_{12} \\ h_{21} & h_{22} \end{pmatrix} * \begin{pmatrix} c_{11} & c_{12} \\ c_{21} & c_{22} \end{pmatrix} \tag{2.11}$$

where (*) denotes the convolution integral:

$$* \to \int dr_3 \, h(r_{13}) \, c(r_{32}). \tag{2.12}$$

Because of the symmetries (2.10) the OZ equation can be diagonalized by a similarity transformation using

$$S = \frac{1}{\sqrt{2}} \begin{pmatrix} 1 & 1 \\ -1 & 1 \end{pmatrix}; \quad S^{-1} = \frac{1}{\sqrt{2}} \begin{pmatrix} 1 & -1 \\ 1 & 1 \end{pmatrix} \tag{2.13}$$

It is easy to verify that, for example:

$$S \begin{pmatrix} h_{11} & h_{12} \\ h_{21} & h_{22} \end{pmatrix} S^{-1} = \begin{pmatrix} h_{11} + h_{12} & 0 \\ 0 & h_{11} - h_{12} \end{pmatrix} \tag{2.14}$$

Therefore, the OZ equation (2.11) becomes a system of two uncoupled, OZ equations. If we define:

$$h^\circ(r) = \frac{1}{2}[h_{11}(r) + h_{12}(r)] \tag{2.15}$$

$$h(r) = h_{11}(r) - h_{12}(r) \tag{2.16}$$

Then we get one "normal" equation,

$$h^\circ(r) - c^\circ(r) = 2\rho \int d\mathbf{r}_1 c^\circ(|\mathbf{r} - \mathbf{r}_1|) h^\circ(\mathbf{r}_1) \tag{2.17}$$

which has the normal boundary conditions for hard spheres and another "special" equation for the charge interactions:

$$h(\mathbf{r}) - c(\mathbf{r}) = \rho \int d\mathbf{r}_1 c(|\mathbf{r} - \mathbf{r}_1|) h(\mathbf{r}_1) \tag{2.18}$$

in which the boundary conditions now have changed. In fact, it is easy to verify that

$$h(r) = 0 \qquad\qquad r < \sigma \tag{2.19}$$

$$c(r) = -\frac{\kappa^2}{\rho r} \qquad\qquad r \geq \sigma \tag{2.20}$$

where

$$\kappa^2 = \frac{8\pi\beta e^2}{\epsilon_0}\rho = \frac{4\pi\beta e^2}{\epsilon}\sum_i \rho_i z_i^2 \tag{2.21}$$

is the Debye screening parameter. Instead of (2.9) we now have:

$$\lim_{\mu \to 0}[1 + \rho h(k)][1 - \rho c(k) + \frac{\kappa^2}{k^2 + \mu^2}] = 1. \tag{2.22}$$

We follow now, step by step, the procedure used in solving the hard sphere case. We write

$$[1 - \rho c(k) + \frac{\kappa^2}{k^2 + \mu^2}] = [1 + \rho Q(k) + \frac{\rho A}{\mu - ik}] \cdot [1 - \rho Q(-k) + \frac{\rho A}{\mu - ik}] \tag{2.23}$$

The inverse Fourier transform of this expression yields:

$$-S(r) + \frac{\kappa^2}{\rho}\frac{e^{-\mu r}}{2\mu} = - Q(r) + A + \rho \int_r^\sigma dr_1\, Q(r_1)\, Q(r_1 - r)$$

$$- \rho \int_r^\sigma dr_1\, Q(r_1)\, A - \rho \int_r^{\sigma+r} dr_1\, A\, Q(r_1 - r)$$

$$+ \rho A^2 \frac{e^{-\mu r}}{2\mu}. \tag{2.24}$$

If we take the limit $\mu \to 0$, then (2.24) requires that

$$\kappa^2 = \rho^2 A^2 \tag{2.25}$$

and, furthermore, $S(r) = 2\pi \int_r^\sigma ds\, s\, c(s)$ is zero at $r = \sigma$, from where we deduce that, due to continuity,

$$Q(\sigma) = 0. \tag{2.26}$$

Consider now the equation for the pair distribution function

$$[1 + \rho h(k)][1 - \rho Q(k) + \frac{\rho A}{\mu + ik}] = \frac{1}{1 - \rho Q(-k) + \frac{\rho A}{\mu - ik}}. \tag{2.27}$$

The Fourier inversion is exactly that of the hard sphere case. We get:

$$J(r) = Q(r) - A + \rho \int_0^\infty dr_1 \, J(r - r_1) \, Q(r_1) - \rho \int_r^\infty dr_1 \, J(r - r_1) \, A \tag{2.28}$$

since now $q(r)$ has become $Q(r) - A$, where A is a constant over the range of r_1 from 0 to ∞. The last term is apparently divergent. Let us write it in the form:

$$\int_r^\infty dr_1 \, J(r - r_1) = - \int_0^r dr_1 \, J(r - r_1) + \int_0^\infty dr_1 \, J(r - r_1). \tag{2.29}$$

but remember that:

$$J(r) = 2\pi \int_r^\infty ds \, s \, h(s) \tag{2.30}$$

so that the last term becomes

$$\rho \int_r^\infty dr_1 J(r_1) = \rho \int_0^\infty dr_1 J(r - r_1) = 2\pi\rho \int_0^\infty dr_1 \int_{r_1}^\infty ds \, s \, h(s)$$

$$= 2\pi\rho \int_0^\infty ds \, s \, h(s) \int_0^s dr_1 = 2\pi\rho \int_0^\infty ds \, s^2 \, h(s) = \frac{-1}{2} \tag{2.31}$$

where the last identity is a consequence of the electroneutrality sum rule for the correlation function $h(r)$ (1.20). Putting it all together yields

$$J(r) = Q(r) - \frac{A}{2} + \rho \int_r^\sigma dr_1 \, J(r - r_1) \, Q(r_1) - \rho A \int_0^r dr_1 \, J(r - r_1) \tag{2.32}$$

Using the condition (2.19), we find the surprisingly simple result

$$J(r) = 2\pi \int_r^\infty ds \, s \, h(s) = b_0 \quad \text{for} \quad r < \sigma \tag{2.33}$$

(we should later see that b_0 is in itself an interesting quantity, namely the excess internal energy E.)
Now (2.32) is

$$b_0 = Q(r) - \frac{A}{2} + \rho b_0 \int_0^\sigma dr_1 \, Q(r_1) - \rho A \, b_0 r \tag{2.34}$$

so that

$$0 = Q'(r) - \rho A b_0 \tag{2.35}$$

and because of the requirement (2.26)

$$Q(r) = \rho A b_0 (r - \sigma) \tag{2.36}$$

so that, taking (2.34) at $r = 0$

$$b_0 = -\rho A b_0 \sigma - \frac{A}{2} - \rho b_0 \frac{\rho A b_0 \sigma^2}{2} \tag{2.37}$$

$$\rho A \sigma = -\frac{2 b_0 \rho \sigma}{(1 + \rho b_0 \sigma)^2} = \kappa \sigma \tag{2.38}$$

to make connection with the Debye- Hückel theory, we define

$$\rho b_0 = -\frac{1}{\sigma + \frac{1}{\Gamma}} \tag{2.39}$$

so that we get, from (2.39)

$$\kappa \sigma = 2 \Gamma \sigma (1 + \sigma \Gamma) \tag{2.40}$$

or

$$(1 + 2 \Gamma \sigma)^2 = (1 + 2 \kappa \sigma)$$

and the physical root for Γ is (a comparison between Γ and κ is given in reference [2])

$$2 \Gamma \sigma = \sqrt{1 + 2 \kappa \sigma} - 1 \tag{2.41}$$

Removing the $\frac{1}{\mu}$ singularity, and taking derivatives of (2.24) we get the direct correlation function:

$$
\begin{aligned}
2\pi \, rc(r) &= -Q'(r) + \rho \int_r^\sigma dr_1 Q'(r_1) Q(r - r_1) + \rho A Q(r) - \frac{\rho A^2}{2} = \\
&= -\rho A b_0 + \rho^2 A b_0 \rho A b_0 \int_r^\sigma dr_1 (r_1 - r - \sigma) - \frac{\rho A^2}{2} + (\rho A)^2 b_0 (r - \sigma) \\
&= \rho A^2 \left[\frac{\rho b_0 \sigma r}{\sigma} + \frac{\rho^2 b_0^2 \sigma^2 r}{2\sigma^2} \right] = \frac{(\rho A \sigma)^2}{\sigma^2} b_0 r (1 + \frac{\rho b_0 r}{2})
\end{aligned} \tag{2.42}
$$

this later expression leads to

$$2\pi c(\sigma) = -\frac{\kappa^2}{\rho_0} \frac{1}{\sigma + \frac{1}{\Gamma}} \left(1 + \frac{\sigma}{2(\sigma + \frac{1}{\Gamma})} \right)$$

This function can be represented by an electrostatic model: It is the electrostatic energy of two spheres around each of the ions as they slide into each other. (see 6.2) We can compute the excess pair correlation function by taking the derivative of (2.32)

$$-2\pi\, r h(r) = Q'(r) - 2\pi\rho \int_0^\sigma dr_1\, Q(r_1)\, (r - r_1)\, h(r - r_1) - \rho A\, J(0) \qquad (2.43)$$

but now $h(r - r_1)$ is zero for $r - r_1 < \sigma$, so that, since $Q'(r)$ and $Q(r)$ are zero for $r > \sigma$

$$2\pi\, r h(r) = \rho A b_0 + 2\pi\rho \int_0^{r-\sigma} dr_1\, Q(r_1)\, (r - r_1)\, h(r - r_1). \qquad (2.44)$$

This equation can be solved by Laplace transformation (which, in the complex plane, is equal to the half-plane Fourier transform (FT))

$$h(s) = \int dr\, e^{-sr}\, r\, h(r). \qquad (2.45)$$

We get

$$h(s) = -\frac{2\Gamma^2 s}{\rho}\, \frac{e^{-s\sigma}}{s^2 + 2s\Gamma + 2\Gamma^2[1 - exp(-s\sigma)]} \qquad (2.46)$$

which should be contrasted to the DH expression $\frac{1}{k^2 + \kappa^2}$ for the FT of the corresponding function. For small concentrations we get, for the more general case, the symmetric expression:

$$h_{ij}(r) \cong \frac{\beta e^2}{\epsilon_0 r}\, \frac{z_i z_j}{(1 + \Gamma\sigma_i)(1 + \Gamma\sigma_j)}\, e^{-(r - \sigma_{ij})\Gamma} \qquad (2.47)$$

which is exponentially decaying, but with a different screening length. In general, however, the function $h(r)$ will be oscillating, modulated by the hard core diameter σ.

Bibliography

[1] R. J. Baxter *Physical Chemistry – An Advanced Treatise,* Edited by H. Eyring, D. Henderson and W. Jost, Vol 8a, p267, Acad. Press, New York, 1968.

[2] D. Wei and L. Blum *J. Phys. Chem.* 91 (1987) 4312.

Chapter 3

Thermodynamic properties

The excess energy can be computed with the help of the pair distribution function [1,10,11] from equation (1.21):

$$\Delta E = 2\pi \int_0^\infty dr\, r^2 \sum_{i,j} \rho_i\, \rho_j\, g_{ij}(r) \frac{z_i z_j e^2}{r\epsilon_0}. \tag{3.1}$$

Using (3.13)-(3.14) and (3.2) manipulations becomes

$$\Delta E = \frac{4\pi e^2}{\epsilon_0} \rho^2 \int_0^\infty dr\, r\, h(r) = \frac{2e^2 \rho^2}{\epsilon_0} b_0 \tag{3.2}$$

and, using the definition of b_0 in (2.33),

$$\Delta E = \frac{2e^2}{\epsilon_0} \rho^2 b_0 = -\frac{e^2 \rho}{\epsilon_0} \frac{1}{\frac{\sigma}{2} + \frac{1}{2\Gamma}} \tag{3.3}$$

where we have also used the relation for b_0. The new screening length 2Γ is clearly that of the MSA. The same picture emerges as in the DH theory: The energy of charging up the system is that of a collection of spherical capacitor of radius $\sigma + \frac{1}{2\Gamma}$. This, in spite of the complicated form of the pair correlation functions. The same simple result is true for the general mixture of arbitrary size ions.

Using formula (1.26) we can compute the free energy excess of the ionic system

$$\Delta A = \frac{1}{\beta} \int_0^\beta d\beta_1\, \Delta E(\beta_1) = \frac{1}{\beta} \int_0^\Gamma d\Gamma' \frac{\partial \beta}{\partial \Gamma'} \Delta E(\Gamma'). \tag{3.4}$$

Now we know that, from equation (2.40)

$$\kappa^2 = \frac{8\pi e^2 \rho \beta}{\epsilon_0} = 4\Gamma^2 (1 + \Gamma \sigma)^2$$

or

$$\frac{\pi e^2 \rho}{\epsilon_0} \frac{\partial \beta}{\partial \Gamma} = \Gamma(1 + \Gamma\sigma)(1 + 2\Gamma\sigma). \tag{3.5}$$

Substituting (3.3) and (3.5) into (3.4) yields:

$$\Delta A = -\frac{2}{\pi\beta} \int_0^\Gamma d\Gamma' \, \Gamma'^2 \, (1 + 2\Gamma'\sigma) = -\frac{1}{\pi\beta} [\frac{2}{3}\Gamma^3 + \sigma\Gamma^4] \tag{3.6}$$

or

$$\Delta A = \Delta E + \frac{\Gamma^3}{3\pi\beta} \tag{3.7}$$

which should be compared to (1.27). Indeed they are the same if we substitute 2Γ for κ using (1.28), (1.29) and

$$\frac{\partial \Gamma}{\partial \rho} = \frac{\pi\beta e^2}{2\epsilon_0} \frac{1}{\Gamma(1 + \Gamma\sigma)(1 + 2\Gamma\sigma)} \tag{3.8}$$

we get the very simple result

$$\phi - 1 = -\frac{\Gamma^3}{3\pi\rho}. \tag{3.9}$$

Finally, the Gibbs free energy per molecule, i. e., the chemical potential $\mu = \frac{\partial G}{\partial n}$, can be calculated:

$$\ln \gamma_\pm = \frac{1}{2}\beta A + \phi - 1 = \frac{1}{2}\beta\Delta E. \tag{3.10}$$

This completes the derivation of these properties of ionic solutions in the MSA. Comparison of the thermodynamic properties to computer simulations show that for low valence and high concentrations, the MSA is comparable to the HNC (Hypernetted Chain Equation) for the activity and osmotic coefficients. For low concentrations and high valence it is not very good [1]. There are a large number of papers in which different ways of correcting this are proposed. A recent approximation, which gives very good results (comparable to the HNC) for 2-2 salts over a range of concentrations varying from 0.00625 M to 2M [1], consists in writing

$$g(r) = A \exp(h^{MSA}(r))S(x) + g^{MSA}(r)(1 - S(x)) \tag{3.11}$$

where $S(x)$ is a switching function (generally linear) which also ensures that the electroneutrality condition (1.20) is satisfied. The nice feature of the MSA is that the simplicity of the results for the equal size case persists for arbitrary mixtures. So, we get to a good approximation:

$$4\Gamma^2 = \frac{4\pi\beta e^2}{\epsilon_0} \sum_i \rho_i \left\{ \frac{z_i}{1 + \Gamma\sigma_i} \right\}^2 \tag{3.12}$$

which is now a higher degree algebraic equation. Often, one can use the equal size equation (1.32) with the mean diameter

$$\bar{\sigma} = \frac{\sum_i \rho_i \sigma_i z_i^2}{\sum_i \rho_i z_i^2} \tag{3.13}$$

as our initial guess for the solution. The excess internal energy is [1]

$$\Delta E = -\frac{e^2}{\epsilon_0} \Gamma \sum_i \frac{\rho_i z_i^2}{1 + \sigma_i \Gamma} \tag{3.14}$$

which again is the sum of the charging energies of a collection of spherical capacitors. The Helmholtz free energy yields, as before

$$\Delta A = \Delta E + \frac{\Gamma^3}{3\pi\beta} \tag{3.15}$$

and, just as before,

$$\phi - 1 = -\frac{\Gamma^3}{3\pi \sum_i \rho_i} \tag{3.16}$$

and

$$\ln \gamma_\pm = \frac{\beta \Delta E}{\sum_i \rho_i} = -\frac{e^2 \Gamma}{\epsilon_0 \sum_i \rho_i} \sum_k \frac{\rho_k z_k^2}{1 + \Gamma \sigma_k}. \tag{3.17}$$

The general solution of the MSA is very useful in many cases to represent the properties of a large variety of eletrolytes and its mixtures, form concentrations ranging from very dilute to almost molten salts [1].

The remarkable fact is that these simple expressions remains true for the case of associating ions [2].

3.1   Variational derivation of the MSA

These results reveal the physical meaning of the MSA solution as an interpolation between two exact lower bounds, the Debye lower bound, effective when $\kappa \to 0$,

$$\beta \Delta E^{MSA,\infty} \simeq -\beta \frac{e^2}{\epsilon} \left[\Gamma \sum_i \frac{\rho_i z_i^2}{1 + \Gamma \sigma_i} \right] \sim -\kappa^2 \qquad \beta \Delta E^{DH,\infty} \sim -\kappa^2 \qquad (3.18)$$

Similarly we get for the entropy

$$\Delta S^{MSA,\infty} \simeq \Gamma^3 \sim -\kappa^{3/2} \qquad \Delta S^{DH,\infty} \sim -\kappa \qquad (3.19)$$

Therefore, in the Onsager limits [3, 4], for the MSA we have

$$\frac{\Delta E^{MSA,\infty}}{\Delta S^{MSA,\infty}} = 0 \qquad (3.20)$$

which are correct. The DH theory also goes to the correct limits, since

$$\frac{\Delta E^{DH,\infty}}{\Delta S^{DH,\infty}} = 0. \qquad (3.21)$$

We can conceive the MSA as a variational problem [5], in which the free energy ΔA is minimal

$$\delta[\Delta A^{MSA}] = 0 \qquad (3.22)$$

which means that, since this is a function of a single parameter

$$\frac{\partial \Delta A^{MSA}}{\partial \Gamma} = 0. \qquad (3.23)$$

is exactly equivalent to the MSA closure. This immediately suggests a rather simple and interesting way of generalizing the MSA: Consider a variational free energy functional, which has the parameter $\lambda_c = \frac{1}{2\Gamma}$, the capacitance length, as the variational parameter. With hindsight we can construct the MSA solution using only dimensional analysis, as follows: (a) Write the excess free energy density in the general form

$$\Delta A(\lambda_c) = \Delta E(\lambda_c) - T \Delta S(\lambda_c) \qquad (3.24)$$

where the expression for the energy density

$$\Delta E = -\frac{e^2}{2\epsilon_0} \sum_i \rho_i \frac{z_i^2}{R_i + \lambda_c} \qquad (3.25)$$

defines the role of this system averaged length scale λ_c, and where we have assumed that the term $S(\lambda_c)$ depends only on λ_c. This is an important assumption, and as a result, and in order to have the correct dimensionality of $[\Delta A(\lambda_c)]$ we write

$$S(\lambda_c) = k\lambda_c^{-3} \tag{3.26}$$

where k is a constant to be determined by adjusting the behavior at low concentrations to the DH picture. The variational equation [5]

$$\frac{\partial \Delta A(\lambda_c)}{\partial \lambda_c} = 0 \tag{3.27}$$

yields

$$\lambda_c^{-4} = -\frac{1}{6A} \frac{e^2}{\epsilon_0 k_B T} \sum_i \rho_i \frac{z_i^2}{(R_i + \lambda_c)^2} \tag{3.28}$$

in the limit $\kappa \to 0$ we expect the DH result to hold. This limit corresponds to $\lambda_c \to \infty$, which yields

$$\lambda_c^{-2} = -\frac{1}{6k} \frac{e^2}{\epsilon_0 k_B T} \sum_i \rho_i z_i^2 \tag{3.29}$$

from where we see that $k = -k_B/24\pi$.

Which again is equivalent to

$$4\Gamma^2 = -\frac{4\pi\beta e^2}{\epsilon} \Gamma \sum_i \rho_i \left(\frac{z_i^*}{1 + \Gamma\sigma_i}\right)^2 \tag{3.30}$$

The generalization to a more general case by minimization of a functional which is identical to the MSA excess free energy ΔA^{MSA} requires that the excess energy be of the form [6, 7]

$$\Delta E = \frac{e^2}{\epsilon} \sum_i \rho_i z_i \frac{X_i - z_i}{\sigma_i}, \tag{3.31}$$

where X_i satisfies an equation of the type

$$\sum_k [\mathcal{M}_{ik}] X_k = z_i, \tag{3.32}$$

where

$$[\mathcal{M}_{ik}] = (1 + \Gamma\sigma_k)\mathcal{J}_{ik} + \mathcal{I}_{ik} \tag{3.33}$$

where $\mathcal{J}_{ik}$ and $\mathcal{I}_{ik}$ are not dependent on Γ. for spherical ions

$$X_i = \frac{z_i - \eta \sigma_i{}^2}{1 + \Gamma \sigma_i}, \tag{3.34}$$

in which the an approximate (the so-called ring sum) free energy ΔA^{MSA} is minimal [5]

$$\delta[\Delta A^{MSA}] = 0. \tag{3.35}$$

Since this is a function of the single parameter Γ, the derivative must be zero

$$\frac{\partial \Delta A^{MSA}}{\partial \Gamma} = 0. \tag{3.36}$$

Differentiation of Eq. (3.23) yields

$$\frac{\partial \Delta A^{MSA}}{\partial \Gamma} = \frac{\partial \Delta E^{MSA}}{\partial \Gamma} - \frac{\partial T \Delta S^{MSA}}{\partial \Gamma}. \tag{3.37}$$

Consider now the following generalization of Eq. (3.32)

$$\sum_k [\mathcal{M}_{ik}] X_k = z_i, \qquad [\mathcal{M}_{ik}] = (1 + \Gamma \sigma_k)\delta_{ik} + T_{ik}, \tag{3.38}$$

where T_{ik} is a matrix which *does not depend on the temperature*. Then, using Eq. (3.31)

$$\frac{\partial \Delta E^{MSA}}{\partial \Gamma} = \sum_i \rho_i \frac{z_i}{\sigma_i} \frac{\partial X_i}{\partial \Gamma}. \tag{3.39}$$

Using the matrix relation

$$\frac{\partial [\mathcal{M}]^{-1}}{\partial \Gamma} = -[\mathcal{M}]^{-1} \frac{\partial [\mathcal{M}]}{\partial \Gamma} [\mathcal{M}]^{-1} \tag{3.40}$$

and Eq. (3.38) we obtain

$$\frac{\partial \Delta E^{MSA}}{\partial \Gamma} = \sum_i \rho_i [X_i]^2. \tag{3.41}$$

Putting it all together yields

$$0 = -\sum_i \frac{\rho_i [eX_i]^2}{\varepsilon} + \frac{\Gamma^2}{\pi}, \tag{3.42}$$

which is an algebraic equation for the new scaling parameter Γ, and is the correct closure equation for the MSA. A similar result was derived by direct calculation in the similar case of a collection of charged sticky spheres by Herrera and Blum [2]. The simplest closure is obtained by ignoring the matrix $[T]$ in Eq. (3.38), which yields a good approximate solution for the MSA for ionic solutions,

$$\sum_i \frac{\rho_i(ez_i)^2}{\varepsilon} \left[\frac{1}{(1 + \sigma_i \Gamma)^2} \right] = \frac{\Gamma^2}{\pi}. \tag{3.43}$$

For low concentrations we get back the DH theory. At infinite coupling we get

$$\kappa \simeq \Gamma^{1/2}, \tag{3.44}$$

which means that our approximations will always satisfy the Onsagerian condition [3],

$$\lim_{\kappa \to \infty} \frac{T \Delta S}{\Delta E} = 0. \tag{3.45}$$

In the last decades the progress of statistical mechanics has opened the possibility of treating quantitatively the effect of ionic interactions at the Mc-Millan Mayer level for clusters [8] [9] [10]. It is possible to include the non ideal contribution in the statistical formulation of the thermodynamic properties of ionic solutions [11] [12] [13]. This can be done combining the concept of ionic association to the evaluation of excess thermodynamic properties.

Bibliography

[1] C. Sanchez-Castro and L. Blum *J. Phys. Chem.* 93 (1989) 7478.

[2] J. N. Herrera and L. Blum *J. Chem. Phys.* 94 (1991) 5077 ; *ibid* 94 (1991) 6190.

[3] L. Onsager *J. Phys. Chem.* 43 (1939) 189.

[4] Y. Rosenfeld and L. Blum *J. Chem. Phys.* 85 (1986) 1556.

[5] D. Chandler and H.C. Andersen *J. Chem. Phys.* 57 (1972) 1930.

[6] E. Velazquez and L. Blum J. Barthel issue of *J. Mol. Fluids.* (In press)

[7] L. Blum, Yu.V. Kalyuzhnyi, O. Bernard, and J.N. Herrera, *J. Phys. Cond. Matter,* 8 (1996) A143.

[8] W.G. MacMillan and J. E. Mayer *J. Chem. Phys.* 13 (1945) 276.

[9] H.L. Friedman and J. C. Rasaiah *J. Chem. Phys.* 48 (1968) 2742.

[10] H.L. Friedman and J. C. Rasaiah *J. Chem. Phys.* 50 (1969) 396.

[11] J. C. Rasaiah *J. Chem. Phys.* 56 (1972) 3071.

[12] L. Blum, M.F. Holovko and I.A. Protsykevytch *J. Stat. Phys.* 84 (1996) 191.

[13] H. R. Corti and R. Fernández Prini *J. Chem. Phys.* 87 (1987) 3052.

Chapter 4

Ion association in the MSA

Ionic association was first discussed by Bjerrrum [1] in conjunction with the DHLL. The extension to the MSA was discussed by [2, 3, 4, 5, 6, 7, 8], but with the assumption that the system was a mixture of ion pairs and free ions which are in chemical equilibrium. It should be noticed that a fundamental distinction exists between the completely dissociated reference state generally used in statistical mechanics, where the excess thermodynamic properties are evaluated taking the ideal gas of ions as the reference system, and the association model in which one postulates the existence of one given state (at least), with a well defined chemical potential. This reference state has to be changed in the regular formalism when the limit of full association is reached. However, a new formalism, developed by Wertheim [9, 10] does include the not only fully associated reference state but also the correct DHLL for the fully associated ions [11, 12]. Implicit in our work is the variational approach to the statistical mechanics of fluids of charged hard objects discussed recently [13]. In the MSA the correlation functions can be considered as variational trial functions of the Free Energy functional. Since, to a very good first approximation, the free energy can be written as the sum of the hard core excluded volume term and an electrostatic term, we can take the diameters of the chemical moeities as different in both terms. This has been done, to a certain extent, in the so called soft MSA [14, 15] with excellent results.

The general theory in which all association processes are properly taken into account using the Wertheim Ornstein Zernike equation (WOZ) [16, 17, 18, 19, 20, 21]has shown that this theory yields very good numerical agreement with computer experiments. The full scaling solution of the binding MSA for (BIMSA)[22, 23] yields the same simple and explicit results in term of a siingle screening parameter Γ. However, this theory includes effects such as the Debye- Falkenhagen effect where the chemical association constant K depends on the ionic strength of the solution.

And it is derived from the same variational principle explained above.

4.1   Chemical equilibrium in non ideal solutions: classical theory

The correct expression of the chemical equilibrium and the corresponding mass action law in solution requires the minimization of the full free enthalpy of the solution, which includes the solvent contribution [24]. We need to minimize the free energy of the solute pair formation equilibrium

$$A + B \Longleftrightarrow AB \tag{4.1}$$

The free energy of the solute is given by

$$A = n_A \mu_A + n_B \mu_B + n_{AB} \mu_{AB} - \Pi V \tag{4.2}$$

where the μ_i 's are the chemical potential of each species and Π is the osmotic pressure and n_i the particle number of species i. For convenience we define

$$\mathcal{A} = \frac{A}{V k_B T} \tag{4.3}$$

Dividing Eq.(4.2 by $V k_B T$ the function to be minimized is

$$\mathcal{A} = \beta(\rho_a \mu_a + \rho_B \mu_B + \rho_{AB} \mu_{AB} - \Pi) \tag{4.4}$$

where ρ_i is the number density and $\beta = 1/k_B T$ If we write the excess $\mathcal{A}$ and Π

$$\mathcal{A} = \mathcal{A}^{id} + \mathcal{A}^{ex} \tag{4.5}$$

and

$$\Pi = \Pi^{id} + \Pi^{ex} = \Pi^{id} + (\phi - 1)\Pi^{id} \tag{4.6}$$

where $\mathcal{A}^{id}$ is the ideal free energy and $\mathcal{A}^{ex}$ is the excess free energy. Similar notation is used for the osmotic pressure Π. The osmotic coefficient is

$$\phi - 1 = \frac{\Pi^{ex} V}{N k_B T} \tag{4.7}$$

We have

$$\mu_i = \mu_i^o + k_B T \ln \rho_i + k_B T \ln y_i = \mu^{id} + \mu^{ex} \tag{4.8}$$

we have then

$$\mathcal{A}^{id} = \rho_a \mu_a^{id} \beta + \rho_B \mu_B^{id} \beta + \rho_{AB} \mu_{AB}^{id} \beta - (\rho_a + \rho_B + \rho_{AB}) \tag{4.9}$$

The excess free energy $\mathcal{A}^{ex}$ and osmotic coefficient ϕ can be computed for the Coulomb [21] and hard sphere parts at the MSA level. we have

$$\mu_i = \frac{\partial G}{\partial n_i} = \mu_i^{id} + k_B T \ln y_i \tag{4.10}$$

and

$$\mathcal{A}^{ex} = \sum_i \rho_i \ln y_i + (\phi - 1) \sum_i \rho_i \tag{4.11}$$

Subject to the constraints imposed by material the balance relations

$$0 = \rho_A + \rho_{AB} - \rho_o \equiv \phi_1 \tag{4.12}$$

$$0 = \rho_B + \rho_{AB} - \rho_o \equiv \phi_2 \tag{4.13}$$

where ρ_o is the initial electrolyte concentration. The minimum of $\mathcal{A}$ Eq. (4.4) subject to the constraints Eq.(4.12) (4.13), is found using the method of the Lagrange multipliers. Consider the function

$$f = \mathcal{A} + \lambda_1 \phi_1 + \lambda_2 \phi_2 \tag{4.14}$$

which is minimized with respect to ρ_A, ρ_B, ρ_{AB} and Γ. This last parameter is the MSA screening parameter. The MSA approximation corresponds to the variational minimum of the electrostatic part of the excess free energy with respect to it [25] and

$$\mathcal{A}^{ex} = \mathcal{A}_{el}^{ex}(\Gamma) + \mathcal{A}_{hs}^{ex} \tag{4.15}$$

The corresponding conditions are

$$\frac{\partial f}{\partial \rho_A} = \frac{\partial f}{\partial \rho_{AB}} = \frac{\partial f}{\partial \rho_B} = \frac{\partial f}{\partial \Gamma} = 0 \tag{4.16}$$

The Lagrange multipliers are

$$-\lambda_1 = \frac{\mu_A^o}{k_B T} + \ln \rho_A + \frac{\partial \mathcal{A}^{ex}}{\partial \rho_A} \tag{4.17}$$

$$-\lambda_2 = \frac{\mu_B^o}{k_B T} + \ln \rho_B + \frac{\partial \mathcal{A}^{ex}}{\partial \rho_B} \tag{4.18}$$

They are essentially the chemical potentials and lead to the mass action law

$$K = \frac{\rho_{AB} f_{AB}}{\rho_A f_A \rho_B f_B} \tag{4.19}$$

where

$$-k_B T \ln K = [\mu^o_{AB} - (\mu^o_A + \mu^o_B)] \tag{4.20}$$

and

$$\ln f_i = \frac{\partial \mathcal{A}^{ex}}{\partial \rho_i} = \ln y_i - (\phi - 1) + \sum_j \rho_j \left(\frac{\partial \ln y_j}{\partial \rho_i} - \frac{\partial \phi}{\partial \rho_i} \right) \tag{4.21}$$

The three last terms in this equation cancel out because of the Gibbs-Duhem relation. We have

$$\ln f_i = \frac{\partial \mathcal{A}^{ex}}{\partial \rho_i} = \ln y_i \tag{4.22}$$

The condition $\partial f / \partial \Gamma = 0$ gives

$$\frac{\partial f}{\partial \Gamma} = 0 = \sum_i \rho_i \frac{\partial^2 \mathcal{A}^{ex}}{\partial \Gamma \partial \rho_i} \tag{4.23}$$

This is equivalent the minimization relation

$$\frac{\partial \mathcal{A}^{ex}}{\partial \Gamma} = 0 \tag{4.24}$$

with $\mathcal{A}^{ex} = \mathcal{A}^{ex}_{el} + \mathcal{A}^{ex}_{hs}$ which can be considered as another definition of the MSA approximation [25]. In any case all excess contributions can be evaluated by an appropriate model like the MSA, taking as the reference state the ideal solution of free ions *and* the pairs at their actual concentrations.

The free energy minimization leads, as required, to the law of mass action , and therefore, we can use any technique to achieve this goal. It should be noticed that except for very simple models, such as the restricted primitive model (RPM) of electrolytes without the hard sphere contributions of the ion pairs, this cannot be done analytically even if explicit expressions of $\mathcal{A}$ are available, but requires the use of computers. From the free energy minimization we obtain the degree of association α.

$$\rho_A = \rho_o(1 - \alpha) \; ; \; \rho_B = \rho_o(1 - \alpha) \; ; \; \rho_{AB} = \rho_o \alpha \tag{4.25}$$

From this free energy, we can separate the excess part by remembering that in this case, the reference state is the partially associated system (3PM). To compare with the 2PM case, we have to remember that the number of particles is not the same in the reference state, e.g. the osmotic coefficient is now given by

$$\phi^{2PM} = \left(1 - \frac{\alpha}{2}\right) \phi^{3PM} \tag{4.26}$$

because the 2PM reference state involves 2 particles while the 3PM reference state involves $2 - \alpha$ particles per mole of dissolved electrolyte. In the same way we get for the activity coefficient

$$\ln y_\pm^{2PM} = \ln(1 - \alpha) + \ln y_\pm^{3PM} \tag{4.27}$$

where $y_{AB}^{3PM} = y_p^{3PM}$ is the activity coefficient of the pairs in the 3PM molar scale. Since the present work deals with simplified models convenient for data fitting, we will take for the excess terms in the free energy average diameters in the case of non restricted primitive model (different ionic radii). They give more tractable expressions, but we have to use two different average diameters: one for the MSA and the other for the hard sphere exclusion part. It has been shown [26] for the electrostatic part that the MSA contribution can be evaluated with an average diameter

$$\sigma_0^{el} = \frac{\sum_i \rho_i \sigma_i z_i^2}{\sum_i \rho_i z_i^2} \tag{4.28}$$

we have then for the electrostatic part of the mean activity coefficient

$$\ln y_\pm = \frac{E^{ex}}{V \rho k_B T} = -\frac{(ze)^2}{\epsilon k_B T} \frac{\Gamma}{1 + \Gamma \sigma_{el}} \tag{4.29}$$

and

$$A_{el}^{ex} = -\frac{\Gamma \kappa_D^2}{4\pi(1 + \Gamma \sigma_{el})} + \frac{\Gamma^3}{3\pi} \tag{4.30}$$

The electrostatic part of the osmotic coefficient is given by

$$\Delta \phi_{el} = -\frac{\Gamma^3}{3\pi \zeta_o} \tag{4.31}$$

with

$$\zeta_n = \sum_k \rho_k (\sigma_k)^n \tag{4.32}$$

and

$$\alpha^2 = \frac{4\pi e^2}{\epsilon k_B T} \tag{4.33}$$

The total osmotic coefficient is split into electrostatic and hard sphere contributions

$$\phi = 1 + \Delta\phi_{el} + \Delta\phi_{hs} \tag{4.34}$$

The condition $\partial \mathcal{A}_{el}^{ex}/\partial\Gamma = 0$, calculated with this value of $\mathcal{A}_{el}^{ex}$ gives

$$\frac{\kappa_D^2}{4} = \Gamma^2 \left(1 + \Gamma\sigma_{el}\right)^2 \tag{4.35}$$

which is identical to the restricted primitive model Γ_o equation for the average diameter

$$\Gamma_o = \left[(1 + 2\kappa_D\sigma_{el})^{1/2} - 1\right]/(2\sigma_{el}) \tag{4.36}$$

The free energy minimization procedure for the chemical equilibrium gives new physical insights for the MSA .

In the same way, the hard sphere contribution can be simplified taking an average diameter which is different from σ_{el}, and is given by an easy to compute expression. At low densities the Percus Yevick compressibility PY_c approximation is an accurate enough to represent the equation of state of hard spheres. We search an expression of the diameter which makes the one component and multicomponent expressions of the PY_c equation of state numerically identical. This is not possible for a single value of σ_{hs}, but rather as an expansion of the form

$$\sigma_{hs} = \sigma_{hs}^0 + \eta\sigma_{hs}^1 + \eta^2\sigma_{hs}^2 + ... \tag{4.37}$$

We are looking for the value of η which will make [1] the PY_c one component osmotic coefficient

$$\phi_{hs} = 1 + \Delta\phi_{hs} = \frac{\beta P}{\rho} = \frac{1 + \eta + \eta^2}{(1 - \eta)^3} \tag{4.38}$$

where

$$\eta = (\pi/6)\sigma_{hs}^3 \sum_i \rho_i \tag{4.39}$$

---

[1] In the mass action law case, we have given the complete Carnahan-Starling expression of the activity coefficient which is valuable for any diameter and any density or packing fraction, here we present a simplified expression with an average diameter and correct for a packing fraction $\eta \ll 0.3$

equal to the PY_c multicomponent term

$$\phi_{hs} = 1 + \Delta\phi_{hs} = \frac{\beta P}{\rho} = \frac{1}{(1 - X_3)} + \frac{3X_1 X_2}{X_o(1 - X_3)^2} + \frac{3X_2^3}{X_o(1 - X_3)^3} \quad (4.40)$$

with

$$X_k = \frac{\pi}{6} \sum_i \rho_i \sigma_i^k \quad (4.41)$$

A naive identification of η and X_3, based on the idea that the excluded volume of individual particles has to be the same, would give

$$\sigma_m = \left[\frac{\sum_i \rho_i \sigma_i^3}{\sum_i \rho_i} \right]^{1/3} \quad (4.42)$$

This approximation takes only into account the singularity as η goes to 1, but gives neither the correct low density nor at high density limits of the osmotic coefficient. For the low volume fractions typical of ionic solutions, identifying the second virial coefficients of the one component and that of the multicomponent models, one finds for the average diameter

$$\sigma_{hs} = \left[\frac{3X_1 X_2 / X_o + X_3}{4X_0} \right]^{1/3} \quad (4.43)$$

This expression constitutes the correct low density approximation for the average hard sphere diameter can be extended to high density [25] For the hard sphere contribution, a convenient and complete expression seems to be a relationship following from the PY theory of hard spheres mixtures [27].

$$\ln y_i^{hs} = -\ln(1 - X_3) + \frac{\sigma_i^3 X_o + 3\sigma_i^2 X_1 + 3\sigma_i X_2}{1 - X_3} + \frac{3\sigma_i^3 X_1 X_2 + \frac{9}{2}\sigma_i^2 X_2^2}{(1 - X_3)^2} + \frac{3\sigma_i^3 X_2^3}{(1 - X_3)^3} \quad (4.44)$$

in the case of an average diameter, (4.44) becomes

$$\ln y_i^{hs} = -\ln(1 - \eta) + \frac{7\eta}{1 - \eta} + \frac{3\eta^2}{(1 - \eta)^2} + \frac{3\eta^3}{(1 - \eta)^3} \quad (4.45)$$

Obviously this method can be extended to mixtures of an arbitrary number of ions just by adding their contributions to the ideal and excess parts of the free energy. The excess parts have been given by expressions (4.44), (4.30), (4.38), (4.40), since the ideal contribution corresponds to a term $\rho_i(\mu_i^o/k_B T + \ln \rho_i)$. If species i is not a reactant protagonist in the association reaction, it will contribute only in the excess free energy part, when one minimizes the free energy with respect to the proportion of pairs, but not in the direct evaluation of concentrations. This model allows to understand the so-called redissociation effects observed in $2 - 2$ electrolytes in water [4] [5]. This effect comes in naturally in the case of a negative excess free energy, but it should be noticed that in the case of positive free energy, which happens for weak electrolytes of large molecular size and high concentrations, the opposite can occur, $i.e.$ reassociation.

Bibliography

[1] N. Bjerrum *Dansk. Vidensk. Felsk. Math. Fys. Medd.* 7 (1926) 9.

[2] W. Ebeling and M. Grigo *Ann. Phys.* 7 (1980) 37.

[3] W. Ebeling and M. Grigo *J. Solution Chem.* 11 (1982) 151.

[4] H. R. Corti and R. Fernández Prini *J. Chem. Phys.* 87 (1987) 3052.

[5] H. R. Corti and R. Fernández Prini *J. Chem. Soc. Faraday Trans.* 86 (1990) 1051.

[6] D. Laría H. R. Corti and R. Fernández Prini *J. Chem. Soc. Faraday Trans.* 86 (1990) 1051.

[7] T. Cartailler, P. Turq, L. Blum and N. Condamine *J. Phys. Chem.* 96 (1992) 6766.

[8] A.C. Tikanen and W.R. Fawcett *Ber. Bunsenges. Phys. Chem.* 100 (1996) 634.

[9] M.S. Wertheim *J. Stat. Phys.* 35 (1984) 19 ; *ibid.* 42 (1986) 459.

[10] M.S. Wertheim *J. Chem. Phys.* 85 (1985) 2929 ; *ibid.* 87 (1987) 7323 ; *ibid.* 88 (1988) 1214.

[11] O. Bernard and L. Blum *Proceedings of the International Conference on Plasmas,* Boston, 1997.

[12] O. Bernard and L. Blum (unpublished).

[13] Y. Rosenfeld and L. Blum, *J. Chem. Phys.* 85 (1986) 1556.

[14] L. Blum and A. H. Narten *J. Chem. Phys.* 56 (1972) 5197 ; A. H. Narten, L. Blum and R.H. Fowler *J. Chem. Phys.* 60 (1974) 3378.

[15] M.J. Gillan *J. Phys. C. Solid State* 7 (1974) L1.

[16] Yu.V. Kalyuzhnyi, M.F. Holovko and A.D.J. Haymet *J. Chem. Phys.* 95 (1991) 9151.

[17] Yu.V. Kalyuzhnyi and V. Vlachy *Chem. Phys. Letters* 215 (1993) 518.

[18] Yu.V. Kalyuzhnyi and M.F. Holovko *Mol. Phys.* 80 (1993) 1165.

[19] M.F. Holovko and Yu. V. Kalyuzhnyi *Mol. Phys.* 73 (1991) 1145.

[20] M.F. Holovko and I.A. Protsykevytch *Mol. Phys.* 90 (1996) 489.

[21] L. Blum, M.F. Holovko and I.A. Protsykevytch *J. Stat. Phys.* 84 (1996) 191.

[22] L. Blum and O. Bernard *J. Stat. Phys.* 79 (1995) 569.

[23] O. Bernard and L. Blum *J. Chem. Phys.* 104 (1996) 4746.

[24] P. Van Rysselberghe *J. Phys. Chem.* 39 (1935) 3.

[25] L. Blum and Y. Rosenfeld *J. Stat. Phys.* 63 (1991) 1177.

[26] C. Sánchez-Castro and L. Blum *J. Phys. Chem.* 93 (1989) 7478.

[27] J. Salacuse and G. Stell *J. Chem. Phys.* 77 (1982) 3714.

Chapter 5

Thermodynamic excess properties of ionic solutions in the primitive MSA

5.1 Introduction

The representation of the departures from ideality in ionic solutions has applications in various domains such as geochemistry, solution chemistry and chemical industry. It is useful for the design of absorption heat pumps or apparatus involved in desalination or nuclear wastes reprocessing.

It is also an interesting information for the study of interfaces in presence of at least one bulk ionic solution.

Because of its wide range of applications, the thermodynamic properties of electrolytes have been the subject of much interest even in recent literature [1, 2, 3]. Certainly, among physical chemists, the most popular expressions have been the Debye-Hückel limiting laws (DHLL), and expressions derived therefrom [4]. Among others, geochemists, have used extensively Pitzer's modifications of DHLL to describe departures from ideality in concentrated ionic mixtures (typically up to 6 mol/kg [5], and up to 10-20 mol/kg, between 0 and 170°C, for solutions of volatile weak electrolytes [6]). Also solubilities of minerals in natural waters can be predicted accurately [7].

Pitzer's treatment is based on the DH theory. It uses the DHLL plus a virial type series correction.

Another theory that is fundamentally connected to the DHLL is the mean spherical approximation (MSA) [8]-[9]. In the DHLL the linearized Poisson Boltzmann equation is solved for a central ion surrounded by a neutralizing ionic cloud. The main simplifying assumption of the DHLL is that the ions in the cloud are point ions. The MSA is the solution of the same linearized Poisson Boltzmann equation but with finite size ions in the cloud. The mathematical solution of the proper boundary conditions of this problem is much more complex. However, simple variational derivations exist nowadays [10].

Calculations of departures from ideality in ionic solutions using the MSA have been published in the past by a number of authors. Effective ionic radii have been determined for the calculation of osmotic coefficients for concentrated salts [11], in solutions up to 1 mol/L [12] and for the computation of activity coefficients in ionic mixtures [13]. In these studies, for a given salt, a unique hard sphere diameter was determined for the whole concentration range. Also, thermodynamic data were fitted with the use of one linearly density-dependent parameter (a hard core size $\sigma(C)$, or dielectric parameter $\varepsilon(C)$), up to 2 mol/L, by least-squares refinement [14]-[16], or quite recently with a non-linearly varying cation size [17] in very concentrated electrolytes.

Parametrization of the thermodynamic properties of pure electrolytes has been obtained [18] with use of density-dependent average diameter and dielectric parameter. Both are ways of including effects originating from the solvent, which do not exist in the primitive model. Obviously, they are not equivalent and they can be extracted from basic statistical mechanics arguments: it has been shown [19] that, for a given repulsive potential, the equivalent hard core diameters are functions of the density and temperature; Adelman has formally shown [20] (Friedman extended his work subsequently [21]) that deviations from pairwise additivity in the potential of average force between ions result in a dielectric parameter that is ion concentration dependent. Lastly, there is experimental evidence [22] for ε being a function of concentration. There are two important thermodynamic quantities that are commonly used to assess departures from ideality of solutions: the osmotic coefficient and activity coefficients. The first coefficient refers to the thermodynamic properties of the solvent while the second one refers to the solute, provided that the reference state is the infinitely dilute solution. These quantities are classic and the reader is referred to other books for their definition [1, 4].

5.2 Strong electrolytes in the MSA

The thermodynamic properties of electrolytes in the primitive MSA have been given elsewhere [23, 24]. For the sake of generality, we will discuss individual ionic excess thermodynamic properties. The single ion activity coefficients for fixed diameters were discussed by several authors [25]-[27]. In all previous work the implicit dependence of the sizes and dielectric constants on the concentration was not taken into account. The discussion below [28] corrects this issue.

The thermodynamic properties can be derived from the Helmholtz energy density A. The excess energy can be split into two terms. One defines

$$\Delta A = \Delta A^{MSA} + \Delta A^{HS} \tag{5.1}$$

where ΔA^{MSA} is the electrostatic contribution that can be calculated in the MSA; ΔA^{HS} is the ion hard sphere contribution.

Electrostatic contribution

Thermodynamic integration (equivalent to Guntelberg charging process) yields an expression for the MSA contribution to A.

One has

$$\beta \Delta A^{MSA} = \beta \Delta E^{MSA} - \int_0^\Gamma d\Gamma' \beta(\Gamma') \frac{\partial \Delta E^{MSA}}{\partial \Gamma'} \tag{5.2}$$

where $\beta = 1/k_B T$. One gets [29]

$$\beta \Delta A^{MSA} = \beta \Delta E^{MSA} + \frac{\Gamma^3}{3\pi} \tag{5.3}$$

The expression for the excess MSA internal energy ΔE^{MSA} (per unit volume) [24] can be rewritten in the following different form.

$$\Delta E^{MSA} = \sum_i \Delta E_i^{MSA} \tag{5.4}$$

with

$$\Delta E_i^{MSA} = -\frac{e^2}{\varepsilon} \rho_i z_i N_i \tag{5.5}$$

with e is the elementary charge and

$$N_i = \frac{\Gamma z_i + \eta \sigma_i}{1 + \Gamma \sigma_i} \tag{5.6}$$

$$\eta = \frac{1}{\Omega} \frac{\pi}{2\Delta} \sum_k \frac{\rho_k \sigma_k z_k}{1 + \Gamma \sigma_k} \tag{5.7}$$

$$\Omega = 1 + \frac{\pi}{2\Delta} \sum_k \frac{\rho_k \sigma_k^3}{1 + \Gamma \sigma_k} \tag{5.8}$$

$$\Delta = 1 - \frac{\pi}{6} \sum_k \rho_k \sigma_k^3 \tag{5.9}$$

The excess osmotic coefficient $\Delta\phi^{MSA}$ is calculated from the thermodynamic relation

$$\Delta\phi^{MSA} = \rho_t \frac{\partial}{\partial \rho_t} \left[\frac{\beta \Delta A^{MSA}}{\rho_t} \right]_{\Gamma=const} \tag{5.10}$$

where

$$\rho_t = \sum_i \rho_i \tag{5.11}$$

In eq 5.10 the derivative is taken at constant Γ, because it is known [10, 29] that

$$\frac{\partial \Delta A^{MSA}}{\partial \Gamma} = 0 \tag{5.12}$$

This result simplifies greatly all the expressions derived hereafter. The single ion activity coefficient reads

$$\Delta \ln y_i^{MSA} = \left[\frac{\partial \beta \Delta A^{MSA}}{\partial \rho_i} \right]_{\Gamma=const} \tag{5.13}$$

Let us define the mean activity coefficient of the mixture by

$$\Delta \ln y_{\pm}^{MSA} = \frac{1}{\rho_t} \sum_i \rho_i \Delta \ln y_i^{MSA} \tag{5.14}$$

It is easy to show [30] that

$$\Delta \ln y_{\pm}^{MSA} = \frac{\beta \Delta A^{MSA}}{\rho_t} + \Delta\phi^{MSA} \tag{5.15}$$

and by differentiation with respect to ρ_t one gets the Gibbs-Duhem relation in the form

$$\frac{\partial \Delta \ln y_{\pm}^{MSA}}{\partial \rho_t} = \frac{\Delta\phi^{MSA}}{\rho_t} + \frac{\partial \Delta\phi^{MSA}}{\partial \rho_t} \tag{5.16}$$

It is observed that one needs only one derivative in all the calculations

$$\left[\frac{\partial \Delta A^{MSA}}{\partial \rho_i}\right]_{\Gamma=const} = \left[\frac{\partial \Delta E^{MSA}}{\partial \rho_i}\right]_{\Gamma=const} \tag{5.17}$$

Remember that both the diameters and the dielectric parameter are a function of the concentration: $\sigma_j(\{\rho_i\}), \varepsilon(\{\rho_i\})$, for i from 1 to n, where n is the number of ions. Using standard implicit function differentiation one gets

$$\Delta \ln y_i^{MSA} = \left[\frac{\partial \beta \Delta E^{MSA}}{\partial \rho_i}\right]_{\Gamma,\rho_k(k\neq i),\sigma_k,\varepsilon} + \sum_j \left[\frac{\partial \beta \Delta E^{MSA}}{\partial \sigma_j}\right]_{\Gamma,\rho_k,\sigma_k(k\neq j),\varepsilon} \left[\frac{\partial \sigma_j}{\partial \rho_i}\right]$$

$$+ \left[\frac{\partial \beta \Delta E^{MSA}}{\partial \varepsilon}\right]_{\Gamma,\rho_k,\sigma_k} \left[\frac{\partial \varepsilon}{\partial \rho_i}\right] \tag{5.18}$$

After some tedious but straightforward algebra one gets

$$\Delta \ln y_i^{MSA} = -\lambda \left[\frac{\Gamma z_i^2}{1+\Gamma\sigma_i} + \eta\sigma_i \left(\frac{2z_i - \eta\sigma_i^2}{1+\Gamma\sigma_i} + \frac{\eta\sigma_i^2}{3}\right)\right]$$

$$+ \sum_j \rho_j q_j \left[\frac{\partial \sigma_j}{\partial \rho_i}\right] + \beta \Delta E^{MSA} \varepsilon \left[\frac{\partial \varepsilon^{-1}}{\partial \rho_i}\right] \tag{5.19}$$

with

$$\lambda = \frac{\beta e^2}{4\pi\varepsilon_0\varepsilon} \tag{5.20}$$

with ε_0 the permittivity of a vacuum. Besides

$$q_j = \lambda \left[\frac{\Gamma^2 z_j^2}{(1+\Gamma\sigma_j)^2} + \eta\frac{\eta\sigma_j^2(2 - \Gamma^2\sigma_j^2) - 2z_j}{(1+\Gamma\sigma_j)^2}\right] \tag{5.21}$$

From this result one finds also, with eq 5.14

$$\Delta \ln y_\pm^{MSA} = \frac{\beta \Delta E^{MSA}}{\rho_t} - \frac{\beta e^2}{\varepsilon}\frac{2}{\pi}\frac{\eta^2}{\rho_t} + \frac{1}{\rho_t}\sum_j \rho_j q_j D(\sigma_j) + \frac{\beta \Delta E^{MSA}}{\rho_t}\varepsilon D(\varepsilon^{-1}) \tag{5.22}$$

with the notation

$$D = \sum_k \rho_k \frac{\partial}{\partial \rho_k} \tag{5.23}$$

and using eqs 5.3, 5.15 and 5.22 one gets

$$\Delta\phi^{MSA} = -\frac{\Gamma^3}{3\pi\rho_t} - \frac{2\lambda}{\pi}\frac{\eta^2}{\rho_t} + \frac{1}{\rho_t}\sum_i \rho_i q_i D(\sigma_i) + \frac{\beta\Delta E^{MSA}}{\rho_t}\varepsilon D(\varepsilon^{-1}) \qquad (5.24)$$

These expressions include the contributions due to the density variation of the diameters and the dielectric constant.

Let us recall that Γ satisfies the closure equation

$$\Gamma^2/(\pi\lambda) = \sum_i \rho_i \left[(z_i - \eta\sigma_i^2)/(1 + \Gamma\sigma_i)\right]^2 \qquad (5.25)$$

Usually, eq 5.25 is easily solved by iteration starting with the initial value $2\Gamma_1 = \kappa$, where κ is the Debye screening parameter

$$\kappa = \left(4\pi\lambda\sum_i \rho_i z_i^2\right)^{1/2} \qquad (5.26)$$

With the mean ionic diameter approximation $\sigma_i = \sigma$ eq 5.25 becomes

$$\Gamma = \left[(1 + 2\kappa\sigma)^{1/2} - 1\right]/(2\sigma) \qquad (5.27)$$

Hard sphere contribution
The hard sphere contribution $\beta\Delta A^{HS}$ is calculated from the Carnahan-Starling approximation [31, 32]. One has that

$$\frac{\pi}{6}\beta\Delta A^{HS} = \left(\frac{X_2^3}{X_3^2} - X_0\right)\ln(1 - X_3) + \frac{3X_1 X_2}{1 - X_3} + \frac{X_2^3}{X_3(1 - X_3)^2} \qquad (5.28)$$

The corresponding contribution to the activity coefficient is calculated in a way similar to eq 5.18. After some algebra one obtains

$$\Delta\ln y_i^{HS} = M_i + \sum_j Q_j \rho_j \frac{\partial\sigma_j}{\partial\rho_i} \qquad (5.29)$$

with

$$M_i = -\ln x + \sigma_i F_1 + \sigma_i^2 F_2 + \sigma_i^3 F_3$$
$$Q_i = F_1 + 2\sigma_i F_2 + 3\sigma_i^2 F_3$$

and

$$F_1 = \frac{3X_2}{x}$$

$$F_2 = \frac{3X_1}{x} + 3\frac{X_2^2}{X_3}\frac{1}{x^2} + 3\frac{X_2^2}{X_3^2}\ln x$$

$$F_3 = \left(X_0 - \frac{X_2^3}{X_3^2}\right)\frac{1}{x} + \frac{3X_1X_2 - X_2^3/X_3^2}{x^2} + 2\frac{X_2^3}{X_3x^3} - 2\frac{X_2^3}{X_3^3}\ln x$$

where

$$X_n = \frac{\pi}{6}\sum_k \rho_k \sigma_k^n$$

$$x = 1 - X_3$$

From eq 5.29 the mean hard sphere activity coefficient can be calculated in the same way as in eq 5.14

$$\Delta \ln y_{\pm}^{HS} = \left(\frac{X_2^3}{X_0 X_3^2} - 1\right)\ln x + \frac{X_3}{x} + \frac{X_2^3(1 + 2X_3 - X_3^2)}{X_0 X_3 x^2} + \frac{3X_1 X_2(2 - X_3)}{X_0 x^2}$$

$$+\frac{1}{\rho_t}\sum_j \rho_j Q_j \mathrm{D}(\sigma_j) \tag{5.30}$$

and since eq 5.15 also holds for the hard sphere part one finds

$$\Delta\phi^{HS} = \frac{X_3}{1 - X_3} + \frac{3X_1 X_2}{X_0(1 - X_3)^2} + X_2^3\frac{X_2^3(3 - X_3)}{X_0(1 - X_3)^3} + \frac{1}{\rho_t}\sum_j \rho_j Q_j \mathrm{D}(\sigma_j) \tag{5.31}$$

5.3 Applications to experiment

5.3.1 Lewis-Randall and McMillan-Mayer description levels

In contrast to Pitzer's work, which is given in molalities (Lewis-Randall theory (LR)), the MSA naturally expresses thermodynamic quantities in terms of concentrations, in the framework of the McMillan-Mayer (MM) theory of solutions [33]. Thus, the data have to be converted from Lewis-Randall to McMillan-Mayer scale for adjusting the model to experiment. The basic ingredients of the LR-to-MM conversion have been given [34] and recently an approximate simple conversion has been tested [35]. The great advantage of this transformation is that it keeps the thermodynamic consistency for the activity and osmotic coefficients, in the sense that these quantities verify the Gibbs-Duhem relation on both LR and MM levels.

It must be underlined that this conversion should be performed in any study using the primitive MSA because, in this framework, the thermodynamic excess functions are calculated at the MM level. The MM framework is characterized by two features: the solvent is regarded as a continuum which manifests itself through its permittivity, and the thermodynamic properties are calculated at constant solvent chemical potential. Although the effect of the conversion is negligible at relatively low concentrations it becomes significant as concentration is increased, typically above 1-2 mol/kg [35].

The relation between LR and MM osmotic coefficients may be expressed [35] as

$$\phi^{(LR)} = \phi^{(MM)} \left(1 - C V_{\pm}\right) \tag{5.32}$$

where $\phi^{(LR)}$ and $\phi^{(MM)}$ are the total osmotic coefficients at the LR and MM level, respectively.

For activity coefficients one may use the following relationship

$$\ln y_i^{(LR)} = \ln y_i^{(MM)} - C V_i \, \phi^{(MM)} \tag{5.33}$$

in which $y_i^{(MM)}$ is given by eq 5.37, V_i is the partial molal volume of species i, and $y_i^{(LR)}$ is the LR activity coefficient in the molarity scale, i.e. [4]

$$y_i^{(LR)} = \gamma_i^{(LR)} V \, d_W \tag{5.34}$$

with $\gamma_i^{(LR)}$ the experimental activity coefficient on the molality scale and d_W the density of pure solvent.

In these expressions C is the total solute concentration

$$C = m/V$$

with m the total solute molality

$$m \equiv \sum_i m_i$$

and V is the volume of solution per mass of solvent in the LR system; $V_{\pm}$ is the mean solute partial molal volume that can be calculated simply [36] from density as

$$V_{\pm} = \frac{M - d'}{d - Cd'} \tag{5.35}$$

where

$$M = \frac{1}{m} \sum_i m_i M_i$$

or

$$M = \frac{1}{m} \sum_S m_S M_S$$

in which the subscript S indicates a salt, m_S its molality, M_S its molar mass, and

$$d' \equiv \left[\frac{\partial d}{\partial C} \right]_{x_i}$$

at constant mole fractions x_i.

For comparison with experiment the quantities are calculated at the MM level as the sum of the electric (MSA) and hard sphere contributions

$$\phi^{(MM)} = 1 + \Delta\phi^{MSA} + \Delta\phi^{HS} \tag{5.36}$$

and

$$\ln y^{(MM)} = \Delta \ln y^{MSA} + \Delta \ln y^{HS} \tag{5.37}$$

5.3.2 Results

Assumptions

The following assumptions have been made [28, 36] about the ion size variation and the permittivity. The anion size was kept constant (equal to its crystallographic value for simple anions or it was adjusted in the case of complex anions) and the diameter of the cation and ε^{-1} were chosen as linear functions of the concentration,

$$\sigma = \sigma^{(0)} + \sigma^{(1)} C_S \tag{5.38}$$

$$\varepsilon^{-1}/\varepsilon_W^{-1} = 1 + \alpha C_S \tag{5.39}$$

where C_S is the molar concentration of the salt, ε_W is the relative permittivity of pure water and $\sigma^{(0)}, \sigma^{(1)}$ and α are parameters.

It is worth noting that common values of $\sigma^{(0)}$, the cation size at infinite dilution which includes hydration water, could be determined [36] for the salts containing the same cation.

It follows from relations 5.38 and 5.39 that

$$D'(\sigma) = \sigma - \sigma^{(0)} \tag{5.40}$$

$$\varepsilon D(\varepsilon^{-1}) = 1 - \varepsilon/\varepsilon_W \tag{5.41}$$

where D is defined by eq 5.23.

Strong Electrolytes

Then the model could be fitted to literature thermodynamic data. Some results for strong electrolytes are shown on Figure 1. Generally the description can be performed [36] to the maximum concentration to which data are available: 16 mol/kg for HCl, 19 mol/kg for LiCl, 20 mol/kg for LiBr,...

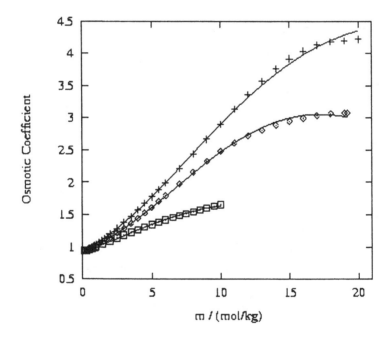

Figure 5.1: LR experimental and calculated osmotic coefficients for LiCl ($\Diamond$), LiBr (+), and LiNO$_3$ ($\square$), as a function of the salt molality.

Associating Electrolytes

Subsequently, the model has been extended [37, 38] to the case of associated electrolytes by using a recent model for associating electrolytes[39]. Unlike the classic chemical model of the ion pair the effect of the pairing association is included in the computation of the MSA screening parameter Γ. Simple formulas for the thermodynamic excess properties have been obtained in terms of this parameter when a new EXP approximation is used. The new formalism based on closures of the Wertheim-Ornstein-Zernike equation (WOZ)[40, 41] does accommodate all association mechanisms (coulombic, covalent and solvation) in one single association parameter, the association constant. The treatment now includes the fraction of particles that are bonded, which is obtained by imposing the chemical equilibrium mass action law. This formalism was shown to be very successful for ionic systems, both in the HNC approximation and MSA [42, 43, 44, 45, 46, 47].

The full solution of the binding MSA for dimer association was discussed elsewhere (BIMSA)[48, 39]. Imposing an exponential closure reminiscent of Bjerrum's approximation [39] for the contact pair distribution function results in simple analytic expressions for the excess thermodynamic quantities.

Figure (5.2) shows the result for two salts up to very high concentration.

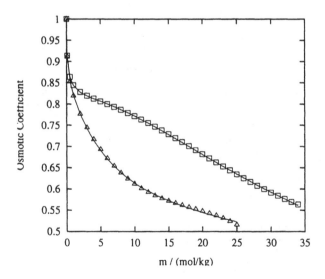

Figure 5.2: Calculated and experimental osmotic coefficients for ammonium nitrate ($\triangle$) and potassium nitrite ($\square$) up to very high concentrations. Notice the sinuous profile in the case of potassium nitrite which is described perfectly by the fit.

Solutions of NaOH and nitric acid, $MgSO_4$,... could be described to high concentration. A description for sulfuric acid has been obtained at low concentration (below 0.1 mol kg^{-1}) with an association constant which compares well with the literature value. Another description has been made for high concentrations (between 6 and 28 mol kg^{-1}) by using a chemically plausible model and a realistic value for the size of the HSO_4^- ion. In most cases the association constant is of the same order of magnitude as the literature value when available.

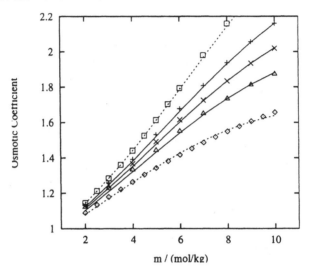

Figure 5.3: LR experimental and calculated osmotic coefficients for the mixture LiCl+LiNO$_3$, for 3 values of x, the fraction of LiNO$_3$: x= 0.3066 ($\triangle$), x= 0.4662 ($\times$) and x= 0.6414 (+), as a function of the total molality. Lines are the calculated curves. The results for the pure salts LiCl and LiNO$_3$ are recalled: x= 0 ($\Diamond$) and x= 1 ($\square$).

Electrolyte Mixtures
Mixtures of strong and associating electrolytes have been described also [36, 38]. For mixed electrolytes a natural extension of eqs 5.38 and 5.39 is

$$\sigma_k = \sigma_k^{(0)} + \sigma_A^{(1)} C_A + \sigma_{B \to A}^{(1)} C_B \qquad (5.42)$$

$$\varepsilon^{-1}/\varepsilon_W^{-1} = 1 + \alpha_A C_A + \alpha_B C_B \qquad (5.43)$$

where a cross term is included, $\sigma_{B \to A}^{(1)}$, to account for the effect of salt B on the size of cation k which belongs to salt A. No cross term is added to the permittivity, for

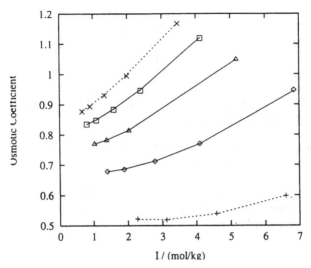

Figure 5.4: LR experimental and calculated osmotic coefficients for the mixture $MgCl_2 + MgSO_4$ for three values of the $MgCl_2$ ionic strength fraction x : $x= 0.246$ ($\Diamond$), $x= 0.493$ ($\triangle$) and $x= 0.752$ ($\square$), as a function of the total ionic strength. The results for the pure salts $MgCl_2$ and $MgSO_4$, from the same reference [49], are also given: $x= 0$ (+) and $x= 1$ ($\times$). Lines are the calculated curves.

which eq 5.39 may be regarded as a first-order expansion of ε^{-1} in powers of the concentration.

Mixtures have been described within this model [38] by using eqs 5.42 and 5.43.

5.3.3 Particular features

The following comments can be made about the results found for the parameters introduced in the model.

- The common values of $\sigma^{(0)}$ are in decreasing order in the series of the alkali cations:
$$\sigma^{(0)}(Li^+) > \sigma^{(0)}(Na^+) > \sigma^{(0)}(K^+)$$
The halides of Rubidium and Cesium ions could be described by taking the crystallographic diameter and imposing $\sigma^{(1)} = 0$ for these ions.

- In nearly all cases, the adjustments have resulted in values for the parameters that have the following satisfying properties:

$$\sigma^{(1)} \leq 0$$

$$\alpha > 0$$

The first relation is interpreted by a decrease in hydration as concentration is increased. The second relation is consistent with theoretical and experimental studies of the dielectric properties of solutions.

- For associating electrolytes the values adjusted for K are of the same order of magnitude as the literature values, when available.

- A nice feature of eqs 5.42 and 5.43 is that no new adjustable parameter is required for mixed electrolytes with a common cation. The parameters of the pure electrolytes will normally suffice:

$$\sigma_k = \sigma_k^{(0)} + \sigma_A^{(1)} C_A + \sigma_B^{(1)} C_B \tag{5.44}$$

where electrolytes A and B share cation k.

Using this procedure, predictions could be made for mixtures with a common cation, without any new parameter. Figures 3 and 4 show the results for such mixtures. The agreement with experimental data was generally excellent.

Bibliography

[1] R.M. Pytkowicz (Ed.) *Activity Coefficients in Electrolyte Solutions*, Vols I and II, CRC Press, Boca Raton, 1979.

[2] A.L. Horvath *Handbook of Aqueous Electrolyte Solutions*, Ellis Horwood, Chichester, 1985.

[3] I.M. Shiah, H.C. Tseng *Fluid Phase Equilibria* 90 (1994) 75, and references therein.

[4] R.A. Robinson, R.H. Stokes *Electrolyte Solutions*, 2nd ed., Butterworths, London, 1959.

[5] K.S. Pitzer and G. Mayorga *J. Phys. Chem.* 77 (1973) 2300.

[6] T.J. Edwards, G. Maurer, J. Newman and J.M. Prausnitz *AIChE J.* 24 (1978) 24.

[7] C.E. Harvie, N. Moller and J.H. Weare *Geochim. Cosmochim. Acta* 48 (1984) 723.

[8] J. K. Percus and G. Yevick, *Phys. Rev.* 110 (1966) 251.

[9] L. Blum and J. S. Høye *J. Phys. Chem.* 81 (1977) 1311.

[10] L. Blum and Y. Rosenfeld *J. Stat. Phys.* 63 (1991) 1177.

[11] S. Watanasiri, M.R. Brulé and L.L. Lee *J. Phys. Chem.* 86 (1982) 292.

[12] W. Ebeling and K. Scherwinski *Z. Phys. Chemie* 264 (1983) 1.

[13] H.R. Corti *J. Phys. Chem.* 91 (1987) 686.

[14] R. Triolo, J.R. Grigera and L. Blum *J. Phys. Chem.* 80 (1976) 1858.

[15] R. Triolo, L. Blum and M.A. Floriano *J. Phys. Chem.* 67 (1976) 5956.

[16] R. Triolo, L. Blum, L. and M.A. Floriano *J. Phys. Chem.* 82 (1978) 1368.

[17] T. Sun, J.L. Lénard and A.S. Teja *J. Phys. Chem.* 98 (1994) 6870.

[18] J.P. Simonin and L. Blum *J. Chem. Soc. Faraday Trans.* 92 (1996) 1533.

[19] H.C. Andersen, D. Chandler and J.D. Weeks *Adv. Chem. Phys.* 34 (1976) 105.

[20] S.A. Adelman *J. Chem. Phys.* 64 (1976) 724.

[21] H.L. Friedman *Kinam* 3A (1981) 101.

[22] J.B. Hasted, D.M. Ritson and C.H. Collie *J. Chem. Phys.* 16 (1948) 1.

[23] L. Blum and J.S. Høye *J. Phys. Chem.* 81 (1977) 1311.

[24] C. Sanchez-Castro and L. Blum *J. Phys. Chem.* 93 (1989) 7478.

[25] J.S. Høye and L. Blum *Mol. Phys. Chem.* 35 (1978) 299.

[26] F. Vericat F. and J.R. Grigera *J. Phys. Chem.* 86 (1982) 1030.

[27] A. Humffray *J. Phys. Chem.* 87 (1983) 5521.

[28] J.P. Simonin, L. Blum and P. Turq *J. Phys. Chem.* 100 (1996) 7704.

[29] L. Blum in *Theoretical Chemistry, Advances and Perspectives*, H. Eyring and D. Henderson eds., vol. 5, Academic Press, New York, 1980.

[30] T. Cartailler, P. Turq, L. Blum and N. Condamine *J. Phys. Chem.* 96 (1992) 6766.

[31] G.A. Mansoori, N.F. Carnahan, K.E. Starling and T.W. Leland *J. Chem. Phys.* 54 (1971) 1523.

[32] J.J. Salacuse and G. Stell *J. Chem. Phys.* 77 (1982) 3714.

[33] W.G. McMillan and J.E. Mayer *J. Chem. Phys.* 13 (1945) 276.

[34] H.L. Friedman *J. Solution Chem.* 1 (1972) 387.

[35] J.P. Simonin *J. Chem. Soc. Faraday Trans.* 92 (1996) 3519.

[36] J.P. Simonin *J. Phys. Chem.* 101 (1997) 4313.

[37] J.P. Simonin, O. Bernard and L. Blum *J. Phys. Chem. B* 102 (1998) 4411.

[38] J.P. Simonin, O. Bernard and L. Blum, *J. Phys. Chem.*, to be published.

[39] O. Bernard and L. Blum *J. Chem. Phys.* 104 (1996) 4746.

[40] M.S. Wertheim *J. Stat. Phys.* 35 (1984) 19; *ibid.* 42 (1986) 459.

[41] M.S. Wertheim *J. Chem. Phys.* 85 (1985) 2929; *ibid.* 87 (1987) 7323 ; *ibid.* 88 (1988) 1214.

[42] Yu.V. Kalyuzhnyi, M.F. Holovko and A.D.J. Haymet *J. Chem. Phys.* 95 (1991) 9151.

[43] Yu.V. Kalyuzhnyi and V. Vlachy *Chem. Phys. Letters* 215 (1993) 518.

[44] Yu.V. Kalyuzhnyi and M.F. Holovko *Mol. Phys.* 80 (1993) 1165.

[45] M.F. Holovko and Yu. V. Kalyuzhnyi *Mol. Phys.* 73 (1991) 1145.

[46] M.F. Holovko and I.A. Protsykevytch *Mol. Phys.* 90 (1996) 489.

[47] L. Blum, M.F. Holovko and I.A. Protsykevytch *J. Stat. Phys.* 84 (1996) 191.

[48] L. Blum and O. Bernard *J.Stat. Phys.* 79 (1995) 569.

[49] Y.C. Wu, R.M. Rush and G. Scatchard *J. Phys. Chem.* 72 (1968) 4048.

Chapter 6

Mathematical background

6.1 Integral and Fourier representation

The solution procedure of the linearized Poisson-Boltzmann equation used above is not suited to include hard core effects of the ions. the most we can do is to give a size to the central ion, but that makes the pair distribution function asymmetric. To include the hard core effects in a symmetric way, we have to change the formalism. We notice, first, that Poisson's equation (1.8) relates the potential $\varphi_i(r)$ to the charge distribution $q_i(r)$. We can formally integrate this equation to yield:

$$\varphi_i(r) = \frac{1}{\epsilon_0} \int d\mathbf{r}_1 \frac{q_i(\mathbf{r}_1)}{|\mathbf{r} - \mathbf{r}_1|} \tag{6.1}$$

which is equivalent to adding up the Coulomb potential at $\mathbf{r}$ produced by all the charges in the system. Clearly, (6.1) must be the same as (1.8). Therefore:

$$\nabla_r^2 \varphi_i(r) = \frac{1}{\epsilon} \int d\mathbf{r}_1 \, q_i(\mathbf{r}_1) \, \nabla_r^2 \left[\frac{1}{|\mathbf{r} - \mathbf{r}_1|} \right] \tag{6.2}$$

For this to be true we must have:

$$\int d\mathbf{r}_1 \, q_i(\mathbf{r}_1) \, \nabla_r^2 \left[\frac{1}{|\mathbf{r} - \mathbf{r}_1|} \right] = -4\pi q_i(\mathbf{r}) \tag{6.3}$$

We introduce the Dirac δ function:

$$\delta(x) = \begin{cases} 0 & \text{if } x \neq 0 \\ \infty & \text{if } x = 0 \end{cases} \tag{6.4}$$

so that

$$\int_{-\infty}^{\infty} dx \, \delta(x) = 1. \tag{6.5}$$

115

And in three dimensions:

$$\delta(\mathbf{r}) = \delta(x)\delta(y)\delta(z) \tag{6.6}$$

which means that if:

$$\nabla_r^2 \left\{ \frac{1}{|\mathbf{r} - \mathbf{r}_1|} \right\} = -4\pi\delta(\mathbf{r} - \mathbf{r_1}) \tag{6.7}$$

Then (6.3) reads:

$$\int d\mathbf{r}_1 \, q_i(\mathbf{r}_1) \left[-4\pi\delta(\mathbf{r} - \mathbf{r_1}) \right] = -4\pi q_i(\mathbf{r}_1) \tag{6.8}$$

which is what we wanted.

We notice that if we multiply (6.7) by $\frac{e z_i}{\epsilon_0}$ we get Poisson's equation for a point charge, the charge density being $q_i(\mathbf{r}) = z_i e \delta(\mathbf{r})$.

We would like to separate the contribution to the potential due to the central particle. In this case (6.1) reads:

$$\varphi_i(r) = \frac{z_i e}{\epsilon_0 r} + \int d\mathbf{r}_1 \, \frac{1}{\epsilon_0 |\mathbf{r} - \mathbf{r}_1|} \sum_j \rho_j \, z_j e \, g_{ij}(\mathbf{r}_1). \tag{6.9}$$

In the DH approximation $g_{ij}(r)$, or also the new quantity

$$h_{ij}(r) = g_{ij}(r) - 1 \tag{6.10}$$

is written

$$h_{ij}(r) = -\beta e z_i \, \varphi_j(r). \tag{6.11}$$

Now substituting (6.11) into (6.9) and using the electroneutrality condition

$$\sum_i \rho_i z_i = 0 \tag{6.12}$$

we get

$$-\beta z_j e \, \varphi_i(r) = -\frac{\beta z_i z_j e}{\epsilon_0 r} - \sum_k \rho_k \int d\mathbf{r}_1 \, \frac{\beta e^2 z_j z_k}{\epsilon_0 |\mathbf{r} - \mathbf{r}_1|} \varphi_i(\mathbf{r}_1) \tag{6.13}$$

or

$$h_{ij}(r) = c_{ij}(r) - \sum_k \rho_k \int d\mathbf{r}_1 \, c_{kj}(|\mathbf{r} - \mathbf{r}_1|) h_{ik}(r_1) \tag{6.14}$$

where we have made the identification

$$c_{ij}(r) = -\frac{\beta e^2}{\epsilon_0} \frac{z_i z_j}{r} = -\beta u_{ij}(r). \tag{6.15}$$

Equation (6.14) can be derived for a much more general class of distributions and is know under the name of Ornstein-Zernike (OZ) equation.

In the discussion of the solution of the OZ equation it will be necessary to unify both descriptions of the Poisson equation: this can be achieved by using the FT technique. Our discussion of the FT will also serve as an introduction to the mathematical techniques used in solving the MSA.

The FT of a function is defined by:

$$\tilde{f}(k) = \int_{-\infty}^{\infty} dx \, e^{ikx} \, f(x). \tag{6.16}$$

The inverse FT is given by:

$$f(x) = \frac{1}{2\pi} \int_{-\infty}^{\infty} dk \, e^{-ikx} \, \tilde{f}(k) \tag{6.17}$$

substituting (6.16) into (6.17):

$$f(x) = \frac{1}{2\pi} \int_{-\infty}^{\infty} dk \, e^{-ikx} \int_{-\infty}^{\infty} dx_1 \, e^{ikx_1} \, f(x_1) \tag{6.18}$$

which is true, since

$$\delta(x - x_1) = \frac{1}{2\pi} \int_{-\infty}^{\infty} dk \, e^{ik(x-x_1)} \tag{6.19}$$

is a representation of Dirac's delta function. In three dimensions:

$$\tilde{F}(k) = \int_{-\infty}^{\infty} dr \, e^{i\mathbf{k} \cdot \mathbf{r}} \, F(\mathbf{r}) \tag{6.20}$$

$$F(\mathbf{r}) = \frac{1}{8\pi^3} \int dk \, e^{-i\mathbf{k} \cdot \mathbf{r}} \, \tilde{F}(k) \tag{6.21}$$

now take the Laplacian of $F(r)$

$$\nabla^2 F(\mathbf{r}) = \frac{1}{8\pi^3} \int dk \, (-k^2) \, e^{-i\mathbf{k} \cdot \mathbf{r}} \, \tilde{F}(\mathbf{k}). \tag{6.22}$$

Consider now the Poisson equation:

$$\nabla^2 \varphi_i(r) = -\frac{4\pi}{\epsilon_0} [q_i(r) - z_i e \delta(r)] \tag{6.23}$$

where we have now included the point charge corresponding to the central ion. In the linearized Bolzmann approximation:

$$\nabla^2 \varphi_i(r) = \kappa^2 \varphi_i(r) + \frac{4\pi}{\epsilon_0} z_i e\, \delta(r). \tag{6.24}$$

We now take the three-dimensional FT:

$$\tilde{\varphi}_i(k) = \int d\mathbf{r}\, e^{i\mathbf{k}\cdot\mathbf{r}}\, \varphi_i(r) \tag{6.25}$$

$$= \frac{4\pi}{k} \int_0^\infty dr\, r\, \sin(kr)\, \varphi_i(r) \tag{6.26}$$

$$\tilde{\varphi}_i(k) = 4\pi \int_0^\infty dr\, \cos(kr) \int_r^\infty ds\, s\, \varphi_i(s). \tag{6.27}$$

Also

$$\int d\mathbf{r}\, e^{i\mathbf{k}\cdot\mathbf{r}}\, \delta(r) = 1 \tag{6.28}$$

and from the Fourier inverse of (6.22) is:

$$\int d\mathbf{r}\, e^{i\mathbf{k}\cdot\mathbf{r}}\, [\nabla^2 \varphi_i(r)] = -k^2\, \tilde{\varphi}(k). \tag{6.29}$$

Putting it all together, the transform of (6.23) is

$$-k^2 \tilde{\varphi}_i(k) = \kappa^2\, \tilde{\varphi}_i(k) + \frac{4\pi z_i e}{\epsilon_0} \tag{6.30}$$

or

$$\tilde{\varphi}_i(k) = -\frac{4\pi z_i e}{\epsilon_0}\, \frac{1}{k^2 + \kappa^2}. \tag{6.31}$$

To compute $\varphi_i(r)$ we need to perform the inverse FT

$$\varphi_i(r) = \frac{1}{8\pi^3} \int d\mathbf{k}\, e^{-i\mathbf{k}\cdot\mathbf{r}}\, \frac{4\pi z_i e}{\epsilon_0}\, \frac{1}{k^2 + \kappa^2}. \tag{6.32}$$

There are two ways of doing this:
(1) Compare to the FT of $\frac{e^{-\kappa r}}{r}$
(2) Use contour integration: to do that we must close a contour around the lower half complex k-plane, where we get a contribution only from the pole located at $k = -i\kappa$.

On other hand, the Fourier Transform of Eq. (6.13) yields

$$\tilde{\varphi}_i = -\frac{z_i e}{\epsilon_0 k^2} - \frac{\kappa^2}{k^2}\tilde{\varphi}_i(k) \tag{6.33}$$

where we have used the property of the Fourier Transform of the convolution of two functions. We remember that the Fourier convolution $f * g$ of two functions $f(x)$ and $g(x)$ is defined by the integral

$$f * g(x) = \int_{-\infty}^{\infty} f(x - \zeta)g(\zeta)d\zeta. \tag{6.34}$$

In terms to FT we have

$$FT(f * g(x)) = FT(f)FT(g). \tag{6.35}$$

and

$$FT(\frac{1}{r}) = \frac{1}{k^2} \tag{6.36}$$

Therefore

$$\tilde{\varphi}_i(k) = -\frac{z_i e}{\epsilon_0}\frac{1}{k^2 + \kappa^2} \tag{6.37}$$

which is the same results that obtained from the differential equation (6.31).

6.2 Direct correlation functions in terms of geometry and electrostatics

We consider an arbitrary mixture of charged hard spheres. The general solution of the MSA [1] yields simple expressions for the thermodynamics and pair correlation functions. The dcf was obtained by Hiroike [2]: For a system of hard spheres of radius $R_i = \sigma_i/2$ charges z_i and number density $\rho_i = N_i/V$, the dcf $c_{ij}(r)$ can be written

$$c_{ij}(r) = c_{ij}^{HS}(r) + c_{ij}^{charge}(r) \tag{6.38}$$

where

$$\gamma_0 = \frac{e^2}{\epsilon_0 k_B T} \tag{6.39}$$

is the Landau length of the system, measuring the relative importance of the electrostatic contributions. When $\gamma_0 = 0$, that is when either the dielectric constant ϵ_0 or the temperature T go to ∞, or when the charges are shutoff by formally letting the electron charge e to be zero, the system corresponds to a neutral uncharged hard sphere mixture for which the dcf $c_{ij}^{HS}(r)$ is that of the Percus-Yevick theory. An important step in constructing a free energy model for the hard sphere mixture begins by casting the known $c_{ij}^{HS}(r)$ in geometric form [3]

$$-c_{ij}^{HS}(r) = \chi^{(3)} \Delta V_{ij}(r) + \chi^{(2)} \Delta S_{ij}(r) + \chi^{(1)} \Delta R_{ij}(r) + \chi^{(0)} \Theta_{ij}(r) \tag{6.40}$$

obeying the MSA closure

$$c_{ij}^{HS}(r > R_i + R_j) = 0 \tag{6.41}$$

For two spheres of radius R_i and R_j at a distance r,

$\Delta V_{ij}(r)$ is the overlap volume,
$\Delta S_{ij}(r)$ is the overlap surface area,
$\Delta R_{ij}(r) = \theta[r - (R_i + R_j)](R_i + R_j - (\text{mean radius of convex}$
$\text{envelope of the union of two spheres}))$
$= \frac{\Delta S_{ij}(r)}{[4\pi(R_i+R_j)]} + \theta[r - (R_i + R_j)](R_i R_j)/(R_i + R_j)$
$\Theta_{ij}(r) = \theta[r - (R_i + R_j)]$

$$\tag{6.42}$$

with

$$\chi^q = \frac{\partial \Phi^{HS}[(\xi_m)]}{\partial \xi_3 \partial \xi_q} \tag{6.43}$$

where

$$\chi^{(0)} = \frac{1}{1 - \xi_3} \tag{6.44}$$

$$\chi^{(1)} = \frac{\xi_2}{[1 - \xi_3]^2} \tag{6.45}$$

$$\chi^{(2)} = \frac{\xi_1}{[1 - \xi_3]^2} + \frac{(1/4\pi)\xi_2^2}{[1 - \xi_3]^3} \tag{6.46}$$

$$\chi^{(3)} = \frac{\xi_0}{[1 - \xi_3]^2} + \frac{2\xi_1\xi_2}{[1 - \xi_3]^3} + (1/4\pi)\xi_2^3 \frac{1}{[1 - \xi_3]^4} \tag{6.47}$$

are the inverse compressibility coefficients in the expansion

$$\chi_i = \sum_q \chi^q[(\xi_m)]R_i^q = \frac{\partial(P/k_B T)}{\partial \rho_i} \tag{6.48}$$

and ξ_q are the fundamental measure variables

$$\xi_q = \sum_q \rho_i R_i^q \tag{6.49}$$

with $R_i^q = V_i, S_i, R_i, 1$ for $q = 3, 2, 1, 0$, respectively

Our first step is to rewrite the dcf of the MSA as written by Hiroike in terms of the geometric and/or electrostatic forms, for the charge part $c_{ij}^{charge}(r)$. We shall present expressions for core overlap configurations, $r \leq (R_i + R_j)$ recalling that the MSA closure is

$$c_{ij}^{charge}(r) = \frac{-z_i z_j}{r} \qquad r > (R_i + R_j) \tag{6.50}$$

After some manipulations Hiroike's dcf can be cast in the form

$$\begin{aligned} c_{ij}^{charge}(r) &= (4/\pi)\eta^2 \Delta V_{ij}(r) \\ &+ 4(R_i R_j)[\eta(X_i + X_j) - (N_i N_j)]\Psi_{ij}(r) \\ &+ 2[N_i X_j - \eta\sigma_i X_j]\Theta_{ij}(r) \end{aligned} \tag{6.51}$$

where $\Psi_{ij}(r)$ is the electrostatic interaction of two charged hard spheres of unit charge smeared on the surface. The spheres are of radius R_i, R_j and they are separated by a distance r. The other parameter of the direct correlation functions are given in terms of a scaling parameter Γ [1], the sizes and charges of the hard spheres, and their concentrations. A new system parameter η, defined by

$$\eta = \sum_i \frac{\rho_i \sigma_i z_i}{1 + \Gamma\sigma_i} \left[\frac{1}{(2/\pi)(1 - \xi_3) + \sum_j \frac{\rho_j \sigma_j^3}{1 + \Gamma\sigma_j}} \right] \tag{6.52}$$

This parameter is related to the symmetry of the solution: In the restricted case, in which all the diameters of the ions are equal, $\eta = 0$. A less restrictive case, in which pair of ions have the same diameter, which however may change from pair to pair, also yields η equal to zero. In terms of these parameters we get

$$X_i = \frac{z_i - \eta\sigma_i^2}{1 + \Gamma\sigma_i} \tag{6.53}$$

$$N_i = -\frac{\Gamma z_i + \eta\sigma_i}{1 + \Gamma\sigma_i} \tag{6.54}$$

The important single parameter of the solution is the capacitance length $(2\Gamma)^{-1} = \lambda_c$, whose role and name will become apparent when we will discuss the thermodynamics of the system in the next section. Before we proceed, however, note that the prefactor of Θ_{ij} in Eq.(6.51) is also symmetric

$$N_j X_i = -[\eta\sigma_i X_j + \eta\sigma_j X_i + \Gamma X_j X_i] \tag{6.55}$$

Unlike the case of the neutral hard spheres, where there is a unique way of writing down the dcf, because of the fact that $\Psi_{ij}(r)$ can be written as a linear combination of the geometric overlap functions, there is no unique way to perform the factorization of the charge dcf. This is in part one of the technical problems that need to be overcome. Specifically, the following relations are true:

$$R_i R_j \Psi_{ij}(R_i, R_j; r) = (R_i + d_i)(R_j + d_j)\Psi_{ij}(R_i + d_i, R_j + d_j; r)$$
$$\begin{cases} -d_i, & r < R_j - R_i \\ -(d_i + d_j)/2, & r > R_j - R_i \end{cases} \tag{6.56}$$

for any positive d_i, d_j,

$$\Psi_{ij}(R_i, R_j) = \frac{\Delta R_{ij}}{R_i R_j}$$
$$= \frac{\Delta S_{ij}}{4\pi R_i R_j(R_i + R_j)} + \frac{\Theta_{ij}}{R_i + R_j} \tag{6.57}$$

and

$$2[N_j X_i - \eta \sigma_i X_j] = \frac{-(2N_j)(2N_i)}{2\Gamma} + \frac{(2\eta\sigma_j)(2\eta\sigma_i)}{2\Gamma} \tag{6.58}$$

We can also write for core overlap configurations

$$c_{ij}^{charge}(r) = (4/\pi)\eta^2 \Delta V_{ij}(r) + \frac{4[\eta(X_i + X_j) - (N_i N_j)]}{4\pi(R_i + R_j)}\Delta S_{ij}(r) + \frac{X_i X_j}{(R_i + R_j)}\Theta_{ij}(r) \tag{6.59}$$

or, if we take into account the boundary condition of the MSA Eq.(6.50) and in view of the Onsager limit, we can write

$$c_{ij}^{charge}(r) = -z_i z_j \Psi_{ij}(r) + (4/\pi)\eta^2 \Delta V_{ij}(r)$$
$$+\frac{4[\eta(X_i + X_j) - (N_i N_j)]}{4\pi R_i R_j(R_i + R_j)}\Delta S_{ij}(r) + \frac{z_i z_j}{R_i R_j}R_{ij}(r) + \frac{[X_i X_j]}{(R_i + R_j)}\Theta_{ij}(r) \tag{6.60}$$

Recalling the results for uncharged hard spheres, we see that independently of the particular decomposition in terms of geometric electrostatic basis "weighted-densities" characterising the geometry of individual particles play the same vital role in the present "charged" case. In the uncharged case we had the scaled particle theory as a guide, and we followed the MSA-compressibility route to the thermodynamics. For the "charge part" we must follow the energy route, so that the result is in a mixed representation, which is more cumbersome.

Bibliography

[1] L. Blum *Mol. Phys.* 30 (1975) 1529

[2] K. Hiroike *Mol. Phys.* 30 (1977) 1195.

[3] Y. Rosenfeld and L. Blum *J. Chem. Phys.* 85 (1986) 1556.

Part III

Specific applications

Chapter 1

Introduction

1.1 Introduction

The study of the interface between two phases which are charged and or conducting is of relevance to a number of systems which occur in nature: colloids, micelles, membranes, solid-solution interfaces in general and metal solution interfaces in particular. These form a bewildering array of systems of enormous complexity. The investigation of the structure of these systems poses considerable difficulties, both experimentally as well as theoretically. The experimental problem is that the interface has 10^{-8} particles relative to the bulk, solid or liquid phases. For this reason one needs a surface specific method, which is able to discriminate between the signal from the surface from the rest. Electrons do not penetrate into solids and for that reason have been used extensively for the ex-situ determination of the surface structure of solids. They must be used in vacuum and that precludes their use in the in-situ study of the liquid solid interface. The study of electrode surfaces removed from the liquid cell under various conditions has provided an enormous wealth of useful data which we will not try to review here. The only way to understand the relation between the ex-situ and in-situ structures is to measure both, something that only nowadays is becoming feasible [1].

1.2 Structure determination

The scanning tunneling microscope (STM) [2] and the atomic force microscope [3] are fascinating new techniques which enable us to see directly the structure of the interfaces. The application of these techniques to electrochemistry is far from trivial, and much progress has been achieved from the initial experiments [4, 5], in which

127

the resolution was relatively poor, until the more recent, rather spectacular pictures [6, 7, 8, 9, 10] with atomic resolution.

While the atomic force microscope (AFM) is a relative newcomer [11, 12], it has some advantages over the STM because it does not measure currents, and therefore does not interfere with the electrochemistry.

The scanning probe microscopes work by moving a sensor needle on the surface to be studied. The sensor tip naturally varies, and is a hard metal needle for the STM , generally tungsten, but Pt-Ir points have also been tried. For the AFM a nonconducting material, such as Si_3N_4 have been employed. The interaction of the tip and the sample is monitored as the tip is moved across the sample by a three dimensional piezoceramic actuator, who provides the scanning motion in the x and y directions, and is moved up and down in the z direction, so as to keep the interaction with the surface constant. This motion is recorded for every position in the x,y plane, and fed into a computer, which then generates 3 dimensional images.

In STM the tips are held 1-2 nanometers above the surface of the electrodes.

If the tip is close enough to the surface there will be a tunneling current, which is an exponential function of the tip to surface distance. This causes very strong variations in current intensity when the tip is moved up or down, the current may typically vary by an order of magnitude when the tip is moved 0.1 nm (1 angstrom) in the z direction. This means that the vertical resolution can be as much as 0.001 nm. The image provided by the STM is really more an electron density map of the surface, and the real positions of the atoms are related to this map through form factors that are known to a certain approximation.

Low resolution scans are certainly less sensitive to these form factors and provide extremely useful morphology information.

In electrochemical environments, the STM is modified to include an integral bipotentiostat which controls independently the voltage of the tip and the surface relative to a given reference electrode. The metal tip is maintained in a potential region in which faradaic processes at the tip are kept to a minimum. In most STM's the working electrode potential is scanned over a certain voltage range, and the tip potential is kept constant. The faradaic current at the tip is kept to a minimum by insulating the sides of the tip, and leaving only the point uncovered. This is a very important detail in electrochemical STM.

The AFM works by a different principle: The tip is mounted on a spring or cantilever, and the force between the tip and the sample is recorded.

In the most popular version of the AFM the deflection of the cantilever is monitored by an optical device consisting of a diode laser focused onto the end of the cantilever and reflected to a split or sectored photodiode. As the cantilever moves the amount of light deflected by the two sectors changes, and generates a current proportional to the deflection of the cantilever. This in turn activates a feedback mechanism that will change the height of the tip so as to bring the cantilever back to its original state. Thus, the height of the tip is adjusted so as to keep the pressure on the surface constant. As in the STM, the height is recorded before the tip is moved to a new x,y position, and the process is re-started. All the information is then fed to a computer, who generates a three dimensional plot.

In electrochemistry, the sample and the tip are immersed in the electrolyte, and it is the electrode that is moved in the solutions. A proper design insures that the proper electrical conections are kept, and no leaks occur during the scanning of the surface.

The tip is one of the most important elements of the AFM. For electrochemical experiments it is made of microfabricated pyramidal Si_3N_4 which is attached to a quarz cantilever. The tips are chemically inert and tolerate 1 N strong acid solutions for long times. Basic media and HF containig solutions cannot be studied with this device.

As in the STM, the AFM does not measure directly the position of the atoms on the surface. It measures the force between the tip and the surface, which is caused by the overlap between the electon clouds of the atoms at the tip and those at the surface. It is clear that the interaction involves not only the atom at the very tip of the needle, but also the neighbors. A detailed theory of the forces in the AFM is not available at the present time.

The STM and the AFM are complementary techniques, and each has advantages of its own. The AFM has the advantage that it does not interfere with the electrochemistry, since the tip is nonconducting and inert. Therefore, measurements can be made even when large faradaic currents are flowing. The AFM can also be used for surfaces other than metals.

The STM provides single atom imaging in electrochemical environment. Steps, single atom defects and other fine details can be seen with the STM. This cannot be done with the AFM, where all of these features appear blurred. The reason for this is that the interaction energy is much lower for the STM, since extremely small currents can be measured, and therefore, it causes less distrubances to the sample. In general, however, there is good agreement between the STM and AFM when the same systems have been observed.

One major drawback of both of these techniques is that although they provide atomic images, there is no way of telling which are the atoms that are seen. Therefore complementary experiments and theoretical interpretations are important to elucidate the structures.

A large number of very interesting experiments have been made recently:

- Direct observation of underpotentially deposited of Cu on Au(111) in the presence of H_2SO_4 [6], provides beautiful pictures of the ordered $\sqrt{3} \times \sqrt{3}$ structures that are attributed to the bisulfate overlayer.

- Surface reconstruction of metals: One of the discoveries of recent times is that the surfaces of many metals even noble metals undergo reconstruction. One of these is Au(111) which undergoes a slow reconstruction to a fishbone structure [13]

- Dynamical processes such as, for example the oxydation of Au(111) surfaces in perchloric acid solutions [14], give direct evidence on the changes in the surface morphology during this process. The formation and dissolution of UPD layers of Pb on Au(111) were also followed by STM [15]: Substantial roughening is observed during the deposition and subsequent stripping of the UPD layer.

- Interactions of small molecules with electrode surfaces can be studied with these techniques: Compression structures of CO on Pt(111) electrode surfaces were exqmined by STM in the presence of H_2SO_4. Several structures were observed, depending on the potential and on the concentration of CO.

- The structure of catalysts is also of considerable interest. The relation of the structure to the chemical reactivity of a catalytic surface is of importance in chemistry. A system thaqt hqs been recently studied is the UPD layer of Bi on Au(111). Chen and Gewirth [12] showed using the AFM that the catalytically active phase was a (2×2) monolayer of Bi on the Au surface, while the more densely packed monolayer was inactive.

1.2.1   X-Ray techniques

X-rays have some unique characteristics for in situ studies in electrochemistry:

1. Hard X-rays of high energy, (around 5 to 10 keV) are of a wavelength comparable to atomic dimensions, and therefore are a probe of the atomic structure of the interface.

2. The hard X-rays have large penetration depths in aqueous solutions (larger than 1mm), so that the electrolyte above the electrode has little effect on the beam.

3. The cross section for X-rays are low, which on one side is an advantage because unlike the STM or even more, the AFM (where the influence of the tip on the

substrate can be noticeable), they will produce a minimal disturbance to the interface. On the other hand, this is a disadvantage, because the signals are also small. The problem of surface sensitive structure determination is that the number of atoms at the surface is roughly 10^{16} while the number of atoms in the bulk is 10^{24}. Therefore we have to measure a signal that in principle is of the order of 10^{-8}compared to the background. However, in all the techniques mentioned above, this goal can be achieved by a combination of extremely powerful X-ray sources and rather specialized detection devices and methods.

It is clear that the Synchrotron radiation sources is the reason that these experiments can be performed. Synchrotron radiation is produced when electrons travelling at almost the speed of light are deflected by a magnetic field perpendicular to the electron beam. From the theory of relativity it can be shown that this will produce a highly polarized, highly collimated and bright beam, which is between 10^5 to 10^{19} times brighter than a conventional X-ray source.

Another important characteristic is the range of energies and wavelengths available from synchrotron radiation. This is very important in EXAFS and XANES experiments, where the energy is scanned rather than the angle.

The study of the electrode interface with X-Rays comprises four powerful methods of structural determination:

- Extended X-Ray absorption fine structure (EXAFS) is a technique in which the sample is submitted to an intense X-ray radiation field, which causes the emission, and subsequent reabsorption of photoelectrons. These electrons are the real probe used by this method: The backscattering gives information not only about the distance of the neighbors of the target atom, but also the near edge structure (X-ray absorption near edge structure, XANES) yields information about state of oxidation and chemical binding of the adsorbates.

- Surface X-ray diffraction is an extension of conventional X-ray diffraction, and is the most accurate and least invasive, (disturbing of the sample) method of structure analysis. It requires crystalline samples of a certain minimum size. However, it can only give accurate measurement of the tangential x-y structure of the electrode surface.

- Low angle X-ray reflectivity (LAXR). The intensity of specularly reflected electromagnetic waves depends on the electron density profile of the surface. The theory of Fresnel [16] is used. At a low angle above the metallic electrode surface, the X-rays are reflected because the refractive index of the metals is slightly lower than 1. Since the interface is rough at the atomic level, there are both reflected and refracted waves that interfere, and produce an oscillating interference pattern.

- Standing waves (XSW) is a technique closely related to LAXR: The incident X-ray beam interferes with a strongly Bragg diffracted or totally reflected beam to create a stationary standing wavefield. The intensity of this field is of the form

$$I = A \sin(az + \varphi) \tag{1.1}$$

where the period a depends on the wavelength of the incident X-rays and the crystal used as electrode, but the phase φ changes with the angle of incidence χ. Thus rocking of the surface makes the stationary waves maxima scan the z direction. This technique is the most difficult experimentally because it has very demanding allignement requirements.[17]

1.2.2   EXAFS

EXAFS measures either the absorption coefficient or the fluorescence from the target atoms, produced by a beam of x-rays of variable energy. This energy is scanned from slightly less to about 1 keV higher than the absorption edge of one of the atoms at the interface to be studied. When the energy crosses the absorption edge, then photoelectrons are emitted, some of which are backscattered and reabsorbed by the target atom. Both the absorption coefficient and the fluorescence produced by the target atom are proportional to the backscattered intensity. For practical reasons electrochemical EXAFS [20, 18, 19] is done with X-rays of about 8 to 10 keV, which means that very light elements are inaccessible to EXAFS, notably oxygen, chlorine, sulphur. Metals like copper, Nickel can be studied using the 1s shells (K edge), and for heavier elements, such as lead, 2p levels (L edge) has to be used. One requisite of SEXAFS is that a certain specific atom has to be adsorbed on the surface as a monolayer. This means that underpotential deposited (UPD) monoatomic films are ideally suited to this technique. In particular the first succesful SEXAFS experiment [20] was performed on a UPD film of Cu on Au(111). The spectra shows a large increase in the signal when the energy crosses the photoionization energy of the target element. The shape of the edge is characteristic of the chemical bonding state of the target atom. The near edge region structure is accounted for by transitions of the photoelectrons to empty states near the Fermi level and to multiple scattering by atoms in the neighborhood of the target atom.[21] Although the theory is in general complicated, the near edge structure yields direct information about the ionization state and bonding of the target atom, which cannot be obtained by any other method. For example the near edge spectra of Cu and of Cu^{++} are shown in

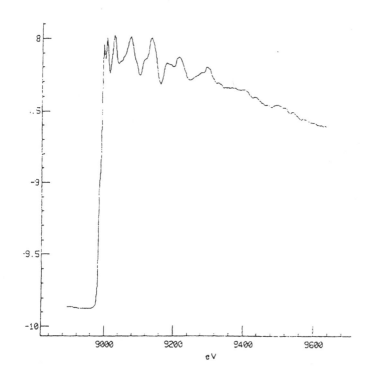

Figure 1.1: Raw spectra of UPD Cu monolayer Au [111].

Figures 1.1 and 1.2 which show the sometimes rather dramatic changes produced by shifts in chemical bonding. When the energy of the photoelectron is bigger than 50 eV, then it ceases to be influenced by the chemistry and the scattering cross section can be explained in terms of the backscattering from the neighboring atoms. This means that although X-rays are used as the primary probe in the experiment, it is the electrons that actually do the structure probing, or in other words, EXAFS is really an electron scattering method, much like EELS, for example, and the X-rays are simply the means of production of the electrons. Therefore, the wavelength of the probe is the wavelength of the photoelectron

$$\lambda = 2\pi/k$$

where the wave number k is given by

$$k = \sqrt{\frac{2m(E - E_0)}{\hbar^2}} \tag{1.2}$$

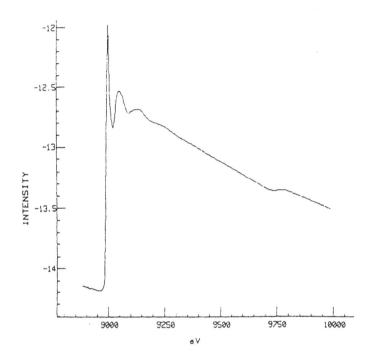

Figure 1.2: Raw spectra of CuSO$_4$.

where $\hbar$ is Planck's constant over 2π, m is the mass of the electron, E is the energy and E_0 is the energy of the edge. In a fluorescence experiment the intensity is given by

$$\chi(k) = \sum_j A_j(k) \sin[2kR_j + \phi_j(k)] \qquad (1.3)$$

where the sum is over the j neighbors of the target atom, R_j is the distance of that neighbor, $\phi_j(k)$ is the phase shift of the photoelectron, which is a function of both the target and the neighbor as well as k. As it happens in electron scattering in general, the structural information depends crucially on the quality and accuracy of both $\phi_j(k)$ and the amplitude $A_j(k)$, which is a complicated but known function of the backscattering neighbor electron density distribution, geometric factors and the Debye-Waller term of the form

$$\exp(-k^2\sigma_j^2)$$

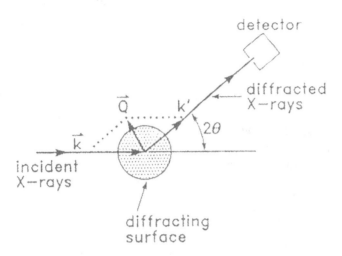

Figure 1.3: Schematic representation of an apparatus for surface diffraction experiment

where σ_j is the mean square displacement of the neighbor from its equilibrium position. This factor implies a strong attenuation of the intensity for atoms thatare loosely bound to the surface. In other words, it will be very hard to see physisorbed molecules of water adsorbed on an electrode by EXAFS. Only chemisorbed atoms will show a sufficiently strong response.

The interpretation of the EXAFS data is therefore not unique, and great care should be exerted in the analysis. But on the other hand it provides information that cannot be obtained in any other way.

1.2.3 Surface diffraction

Figure 1.3 gives a schematic representation of an apparatus for surface diffraction experiment.

The wave vectors of the incident and diffracted beams are **k** and **k'**. Since we only deal with elastic scattering they are both of the same magnitude

$$k = k' = 2\pi/\lambda$$

where λ is the wavelength of the X-rays. The intensity of the diffraction spots (or rods) depends on

$$\mathbf{Q} = \mathbf{k}' - \mathbf{k}$$

The magnitude of Q is given by

$$Q = \frac{4\pi}{\lambda} \sin \theta \tag{1.4}$$

where 2θ is the angle between $\mathbf{k}$ and $\mathbf{k}'$

For crystalline bulk materials the X-ray diffraction pattern consists of a series of spots or Bragg peaks [22], with an intensity given by

$$I(\mathbf{Q}) = 2Re\left[\sum_{\mathbf{n}} f(\mathbf{n})e^{R_{\mathbf{n}} \cdot \mathbf{Q}} \right] \tag{1.5}$$

where $\mathbf{Q}$ is a function of Q and the sample orientation, $f(\mathbf{n})$ is the form factor of atom $\mathbf{n}$ and $R_{\mathbf{n}}$ its position given by the angles χ and ϕ.

In Eq.(1.5) f_i is the form factor for atom i, which is proportional to the square of the atomic number, or the number of electrons. This means that the cross section for oxygen is 64 times larger than that of hydrogen, and for example, lead, with an atomic number of almost 100, has a cross section of about 100,000 times that of hydrogen. This illustrates very dramatically the fact that only heavy elements are accessible to surface diffraction by X-rays.

The peak positions are at the projections of the positions of the points of the reciprocal lattice of the crystal that is studied. The peak positions determine the unit cell's size and symmetry, while a careful analysis of the intensities of the different reflections yields the positions of the atoms within the unit cell.

When a highly perfect crystal pattern is observed, very faint streaks between the diffraction spots can be observed, which are the so called diffraction rods. The observation of these diffraction rods is facilitated by the glancing angle technique, which consists in shining the X-ray beam at the critical angle at which total reflection occurs (remember that the index of refraction of metals to X-rays is lower than 1, so that for small angles of the order of 0.1 degrees total reflection occurs). At this angle there is a strong enhancement of the refracted beam that will travel along the surface. This is known as the surface enhancement.
The truncation rods therefore are truly two dimensional scattering patterns which

contain 2 dimensional sums rather than 3 dimensional sums as in the bulk case.

The intensity of the trunction rods is influenced by the structure of the crystal near the surface as well as that of the overlayer, or fluid near the electrode. The measurement of the truncation rods permits the determination of the structure of the adsorbed atoms relative to the positions of the substrate atoms, and also the rough features of the density distribution above the electrode.

Experimental results are available for the diffraction rods and structure of an overlayer of Pb on Ag(111)[23, 24]. Since these measurements are very accurate it is also possible to measure the compression of the UPD overlayer of Pb as the potential is changed[25, 26]. The compressibility is certainly related to the electrosorption valency, discussed in another section of this book.

1.2.4 Other in-situ methods

Other in-situ techniques give information that is thermodynamic in nature since it comprises the average over a number of atoms. One technique that has been established recently is the quarz microbalance[27]: this instrument can measure small changes in the mass of a metallic electrode that is attached to a quarz oscillator. It has the disatdvantage of not being able to de-couple the dynamics of the inner double layer from that of the diffuse double layer. The electrosorption valency, for example can be calculated directly,by measuring the mass deposited at the electrode and the amount of charge from voltammogram. The interpretation of the results of this instrument requires electrode surfaces that have large molecularly smooth regions. The spectroscopic methods using ultraviolet, visible or Raman spectroscopy [28] are very useful in situ probes because a large number of organic molecules can be studied . Interesting information about changes in bonding, and symmetry can be extracted. The optical spectroscopic methods do not require special installations such as the synchrotron, and are most useful for complex molecular species. The techniques are the surface enhanced Raman ,surface infrared spectroscopy, second harmonic generation , which permits to discriminate between different geometries of the adsorbates on single crystal surfaces.

Amongst the optical techniques there are also the more traditional methods such as the ellipsometry, electroreflectance and particularly, surface plasmons , where experimental and theoretical advances have made it possible to offer a picture of the surface electronic states of the metal in some selected cases, such as the silver (111) phase. We should mention here the measurement of image potential induced surface states by electroreflectance spectroscopy . In this case, besides the normal surface states which arise from the termination of the crystal lattice, there are discrete states due to the existence of an image potential for charges near the conducting interface.

A method that has yielded very interesting information about the structure and interactions in the diffuse part of the double layer is the direct measurement of forces between colloidal particles [29].The forces between two mica plates are measured directly in the presence of different solutions: These forces show pronounced oscillations of a period similar to the dimensions of the molecules enclosed between the plates. And last, but certainly not least, there is a very extensive and important literature on the differential capacitance of solutions near either solid (polycrystalline or single crystal) or liquid (mercury) electrodes which we will not try to cover . We should mention the recent work on the influence of the crystallographic orientation of silver on the potential of zero charge of the electrodes, in which a detailed mapping of the influence of the crystal face on the differential capacitance of the inner layer is made [30].

1.3   Theory

The complexity of the system described by the experimental methods defies any simple theoretical interpretation. Yet these are needed for the understanding of what is actually going on at the charged interface. It is clear that the simplest theory should include in the discussion two kinds of forces : the long ranged Coulomb forces and the short ranged forces that are at the origin of the chemical bonds and are also responsible of the repulsion between atomic cores. There are important quantum effects at the interface due to the quantum nature of the electrons in a metal[31]. These effects are very difficult to compute in a proper way, and in most theoretical discussions only very sketchy models of the liquid side of the interface is discussed when attempting to describe the metal side of the interface.

For this reason we have organized the theoretical discussion starting with very simple model systems about which a lot is known, and to systems which are much more realistic but difficult to handle . The emphasis of the theoretical treatment will be on the structure functions, or distribution functions $\rho(1), \rho(1, 2), ...$ which give the probability of finding an ion(s) or solvent molecule(s) at specified position(s) near the interface. The properties of the interface can then be calculated from these distribution functions. One of the very interesting theoretical developments of recent years has been the exactly solvable model developed by Jancovici, Cornu and co-workers . This is a two dimensional model at a particular value of the reduced temperature, and is particularly useful to elucidate the subtle properties of the long ranged Coulomb forces. For the non primitive model with solvent molecules there is a one dimensional exactly solvable model . Exactly solvable models serve as benchmarks for approximate theories and to test exact and general sum rules.

We start with the simplest model of the interface, which consists of a smooth charged hard wall near a ionic solution that is represented by a collection of charged hard spheres, all embedded in a continuum of dielectric constant ϵ. This system is fairly well understood when the density and coupling parameters are low. Then we replace the continuum solvent by a molecular model of the solvent. The simplest of these is the hard sphere with a point dipole[32], which can be treated analytically in some simple cases. More elaborate models of the solvent introduce complications in the numerical discussions. A recently proposed model of ionic solutions uses a solvent model with tetrahedrally coordinated sticky sites. This model is still analytically solvable. More realistic models of the solvent, typically water, can be studied by computer simulations, which however is very difficult for charged interfaces. The full quantum mechanical treatment of the metal surface does not seem feasible at present. The jellium model is a simple alternative for the discussion of the thermodynamic and also kinetic properties of the smooth interface [33, 34, 35, 36, 37, 38, 39, 40].

1.3.1 Exact results and theorems

There are a number of exact sum rules that the density profiles neqr electrodes have to satisfy. One set of these rules is due to the special long range nature of the Coulomb forces, which give rise to the screening of the charges in conducting media. The second set of sur rules originates from force balance requirements, and are the dynamic sum rules.

The screening sum rules

The screening sum rules, are specific to Coulomb forces. Because of the very long range of the electrostatic forces, the stability of the system requires that all charges surround themselves with a neutralizing cloud. The surface charge satisfies the electroneutrality condition

$$-q_s = \int_0^\infty dz \sum_{i=1}^m e_i \rho_i(z) = \epsilon E_0/4\pi \tag{1.6}$$

where E_0 is the external or applied field In homogeneous systems of molecules interacting with Coulomb forces the screening of charges and multipoles by the conducting media is intuitive because of the isotropy of the system. In a homogeneous solution every charge is surrounded by an ionic cloud of exactly the opposite charge. But also every dipole (or for that matter any arbitrary neutral charge distribution) is surrounded by a charged cloud that has a dipole moment (charge distribution) exactly opposite that of the original dipole, and in general, one can show [41, 42, 43] for any charge distribution in an homogeneous system. The fact is that it is also true in the inhomogeneous case, which is not intuitively obvious, and has been confirmed by the beautiful work of Jancovici [44, 45, 46, 47, 48, 49], for an exactly solvable two dimensional model. The demonstration of these theorems is based on the Born-Green-Yvon hierarchy (BGY) and the an assumption on the clustering of the correlation functions. In electrically neutral systems any fixed arrangement of charges is screened by the mobile charges of the system. In homogeneous bulk phase this is an intuitively natural fact, because if the long ranged Coulomb forces would not be screened then the partition function would not exist (it would diverge), and matter would not be stable[50]. This is expressed by the fact that the charge distribution around a given charge e is of equal value but opposite sign. In the homogeneous bulk phase this is a natural fact:

$$-e_i = \int d2 \sum_j e_j \rho_j h_{ij}(1,2) \qquad (1.7)$$

Rotational invariance in bulk fluids requires that not only charges but also multipole of arbitrary order should be screened by the mobile charges of the media [41, 42]. This fact is much less intuitive in the neighborhood of charged objects, in particular in the neighborhood of a charged electrode. However the theorems hold and in classical mechanics, at least, perfect screening of all multipoles occurs, in the homogeneous or inhomogeneous systems. However, perfect screening of all multipoles does not occur in quantum systems or in systems out of equilibrium [51]. As a consequence of the screening the second moment of the pair distribution function must be normalized. This is the Stillinger-Lovett [52] moment relation . Outhwaite[53],has shown that it can be written as a normalization condition for the electrostatic potential

$$\psi_i(r) = 1/\epsilon \left[e_i/r - \sum_j e_j \rho_j \int dr_1 \frac{h_{ij}(r_1)}{(|\mathbf{r} - \mathbf{r}_1|)} \right] \qquad (1.8)$$

which satisfies the sum rule

$$1 = 1/kT \sum_j e_j \rho_j \int d\mathbf{r}_1 \psi_j(1) \qquad (1.9)$$

Carnie and Chan [54] have shown that this normalization condition is also valid for the inhomogeneous systems of charged particles.

For flat hard electrode surfaces there are number of other sum rules. A relevant sum rule for the calculation of density profiles in the electric double layer is the dipole rule [55]

$$kT\partial\ln\rho_i(1)/\partial E_0 = \int d2 \sum_j e_j\rho_j(2)h_{ij}(1,2)(z_1 - z_2) \tag{1.10}$$

where E_0 is the bare field at the electrode surface. The differential capacity, which is defined by

$$C_d = \partial q_s/\partial\Delta\phi \tag{1.11}$$

where q_s is the surface charge, $q_s = \frac{E_0\epsilon}{4\pi}$, and $\Delta\phi$ is the potential drop, satisfies the sum rule

$$1/C_d = \frac{8\pi^2}{\epsilon^2 S} \int d1 d2 \sum_{ij} e_i e_j \rho_i(1)\rho_j(2)h_{ij}(1,2)(z_1 - z_2)^2 \tag{1.12}$$

Dynamic Sum rules

These sum rules are derived from balance of forces considerations. Systems interacting with conservative forces must satisfy momentum conservation and force balance. This apparently trivial requirement is not satisfied by some of the approximate theories used in the description of the electrode interface. We consider a system which is limited by an arbitrary surface, which could be planar, but also a rough surface which is planar in the average. We include single crystal metal surfaces, but also macroscopically smooth, but microscopically rough interfaces. The precise mathematical requirement is that there is a prism with an arbitrarily large cross section area S, and height L (the volume $V = SL$), such that the force through the walls parallel to z is of $O(S^{1-\delta})$, where $\delta \to 0$, as $S \to \infty$. In our notation $\mathbf{r} \equiv (x, y, z)$, x, y are the coordinates in the electrode plane, and z is normal to the electrode plane.

We integrate the force balance equation in the volume of a prism of the same section S but smaller height $L_1 < L$, summing over all ionic and neutral species i, For central forces in the average over the surface $\mathbf{S}$ the pair interactions are cancelled out exactly. At the plane L_1 these forces will contribute to the bulk pressure P. The statement of dynamic balance is that

$$P = \bar{p} \tag{1.13}$$

where $\bar{p}$ is the normal pressure at the electrode wall:

$$\bar{p} = -kT \sum_{i=1}^{m} \frac{1}{S} \int dxdy \int_S dz \frac{\partial w_i(\mathbf{r})}{\partial z} \rho_i(\mathbf{r}) \qquad (1.14)$$

The interaction of the molecules in the fluid and the electrode are represented by $w_i(\mathbf{r})$. This function is in the most general case the sum of two contributions, a Coulomb or electrostatic term plus a non electrostatic, covalent term such as van der Waals, hard core, etc.

$$w_i(\mathbf{r}) = w_i^{cov}(\mathbf{r}) + w_i^{es}(\mathbf{r}) \qquad (1.15)$$

Consider different situations:

1. The flat electrode face a primitive model (continuum dielectric) electrolyte. In this case

$$w_i(\mathbf{r}) = w_i^{hc}(\mathbf{r}) + w_i^{es}(\mathbf{r}) \qquad (1.16)$$

The hard core potential is best represented by it force

$$\frac{\partial w_i^{hc}(z)}{\partial z} = -k_B T \delta(\sigma_i/2 - z) \qquad (1.17)$$

while the electrostatic contribution is

$$w_i^{es}(z) = -e_i E_0 z \qquad (1.18)$$

where e_i is the charge of i, E^0 is the bare electric field. Using the electroneutrality relation Eq.(1.48) we immediately get

$$P = k_B T \sum_{i=1}^{m} \rho_i(\sigma_i/2) - \frac{\epsilon}{8\pi}[E_0]^2 \qquad (1.19)$$

which expresses the fact that the total pressure must be equal to the kinetic term due to the collisions of the molecules at the wall minus the attractive electrostatic contribution of a planar capacitor with charge density $q_S = \frac{\epsilon E_0}{4\pi}$.

2. The flat electrode face a non primitive model electrolyte. If the solvent consists of hard spherical neutral molecules with a dipole (or higher multipoles), there will be no net force since the dipoles interact with the gradient of the

applied field $\nabla \mathbf{E_0}$, which in this case is zero. Therefore, only the hard repulsive interactions count

$$\frac{\partial w_s^{hc}(z)}{\partial z} = -k_B T \delta(\sigma_s/2 - z) \tag{1.20}$$

and we get now

$$P = k_B T \left[\sum_{i=1}^m \rho_i(\sigma_i/2) + \rho_s(\sigma_s/2) \right] - \frac{1}{8\pi}[E_0]^2 \tag{1.21}$$

Notice that now the dielectric constant has disappeared from this relation. This means that the electrostatic contribution in a solvent of high dielectric constant like water, is now much smaller than in the primitive model, and that the hard core term plays a much larger role in the makeup of the concentration profile near the electrode.

3. The rough or structured electrode near a primitive electrode. The situation is now more complicated since the charge distribution at the electrode surface will not be uniform, and therefore, both the contact density (for hard surfaces as well as for soft surfaces), will be functions of z as well as the position on the surface x, y. There will be a simple relation only for average quantities such as the average contact density near a hard plane,

$$\bar{\rho}_i(0) = 1/S \int dx_1 dy_1 \rho_i(x_1, y_1, z_s) \tag{1.22}$$

It is clear that the electrostatic forces along the surface are of vanishing magnitude for a random rough surface, or zero for a periodic crystal surface.

Using Poisson's equation

$$\nabla^2 \psi(1) = \frac{-4\pi}{\epsilon} \sum_{i=1}^m e_i \rho_i(1) = \vec{\nabla}_1 \cdot \mathbf{E}(1) \tag{1.23}$$

and integrating by parts, we get

$$\int d1 \sum_{i=1}^m e_i \rho_i(1)[E_z(1)] = -\epsilon/8\pi \int_S dx_1 dy_1 \int_{z_s(x_1,y_1)}^{L_1} dz_1 \frac{\partial [E_z(1)^2]}{\partial z}$$

$$+\epsilon/4\pi \int_V dx_1 dy_1 dz_1 \left[E_z(1) \frac{\partial E_x(1)}{\partial x_1} + E_z(1) \frac{\partial E_y(1)}{\partial y_1} \right] \tag{1.24}$$

The second term of the right hand side is zero: For a periodic interface in the x and y directions, if we take S to be the surface of a unit cell, the terms like

$$\frac{\partial E_x(1)}{\partial x_1}$$

will be of equal magnitude but of opposite sign for neighboring cells. For the general random interface we conjecture that this term is finite: then in the limit $S \to \infty$ the contribution vanishes. We have

$$1/S \int d1 \sum_{i=1}^{m} e_i \rho_i(1) E_z(1) = -\epsilon/8\pi < [E_z(1)]^2 >_S \qquad (1.25)$$

where the average square field in the z direction is

$$< [E_z(1)]^2 >_S = 1/S \int_S dx_1 dy_1 E_z^2(x_1, y_1, z_s) \qquad (1.26)$$

The other single particle term containig the short range interactions between the molecules and ions and the wall, yields

$$< \rho_i(1) \frac{\partial w_i^{cov}(1)}{\partial z_1} >_S = 1/S \int_S dx_1 dy_1 \rho_i(1) \nabla_1 w_i^{cov}(1) \qquad (1.27)$$

where

$$< \rho_i(1) \frac{\partial w_i^{cov}(1)}{\partial z_1} >_S = 1/S \int_S dx_1 dy_1 \int_{z_s(x_1,y_1)}^{L_1} dz_1 \rho_i(1) \frac{\partial w_i^{cov}(1)}{\partial z_1} \qquad (1.28)$$

Putting it all together yields the general contact theorem for a planar on the average, but not necessarily smooth, surface

$$P = k_B T \sum_{i=1}^{m} \bar{\rho}_i(\sigma_i/2)) - \epsilon/8\pi < [E_z(1)]^2 >_S$$

$$- \sum_{i=1}^{m} < \rho_i(1) \frac{\partial w_i^{cov}(1)}{\partial z_1} >_S \qquad (1.29)$$

This theorem is a generalization of the previously derived contact theorems to the realistic case of non smooth electrode surfaces. It contains the previous results as particular cases.

When the walls are soft, then

$$P = -\sum_{i=1}^{m} < \rho_i(1) \frac{\partial w_i^{cov}(1)}{\partial z_1} >_S -\epsilon/8\pi < [E_z(1)]^2 >_S \qquad (1.30)$$

For a surface with an array of sticky adsorption sites, such as in the case of the sticky site model, (SSM) model discussed elsewhere [56, 57, 58] , the adsorption potential has the form

$$e^{-\beta u_a(r)} = 1 + \lambda_a(\mathbf{R})\delta(z) \qquad (1.31)$$

with

$$\lambda_a(\mathbf{R}) = \sum_{n_1,n_2}^{M} \lambda_a\delta(\mathbf{R} - n_1\mathbf{a_1} - n_2\mathbf{a_2}) \qquad (1.32)$$

Here $\mathbf{R} = x, y$ is the position at the electrode surface, and z the distance to the contact plane, which is at a distance $\sigma/2$ from the electrode. In Eq.(1.32), n_1, n_2 are entire numbers, there are M sites on the electrode of area S , and $\mathbf{a_1}, \mathbf{a_2}$ are the lattice vectors of the adsorption sites on the surface. The parameter λ_a represents the fugacity of an adsorbed atom of species a. Define now the regular part of the density function

$$y_i(1) = (\rho_i(1)/\rho_i)e^{\beta w_i^{cov}(1)} \qquad (1.33)$$

Replacing into the general contact theorem Eq.(1.29) gives [55]

$$P = k_B T \sum_{i=1}^{m} \bar{\rho}_i(\sigma_i/2) - \epsilon/8\pi < [E_z(1)]^2 >_S + \frac{M\lambda_a}{S} \sum_{i=1}^{m} < \frac{\partial y_i(1)}{\partial z_1} > \rho_i \quad (1.34)$$

4. The rough electrode near a non primitive (for example with a solvent of dipolar hard spheres) electrolyte. Now we have to include the effect of electric field gradients, which are not zero near the electrode. The total electrostatic force is [60]

$$-\frac{\partial w_\alpha^{es}(\mathbf{r})}{\partial \mathbf{r}} = e_\alpha \mathbf{E}_0 + \mu_\alpha.(\vec{\nabla}\mathbf{E}_0) + (1/6)q_\alpha : (\vec{\nabla}\vec{\nabla}\mathbf{E}_0) \qquad (1.35)$$

where μ_α is the dipole moment of α, q_α its quadrupole moment, and so on. We remark that now the single particle density $\rho_s(1)$ is not only a function of $\mathbf{r}_1$, but also of the orientation of the molecules with respect to the electrode, which in the case of the linear dipoles is given by θ_1, ϕ_1. Therefore we expand

$$\rho_s(1) = \sum_{l,m} \rho_{s,m}^{\ell} Y_m^{\ell}(\theta_1, \phi_1) = \rho_{s,0}^0 + \sum_m \rho_{s,m}^1 Y_m^1 + ... \qquad (1.36)$$

The dipole contribution to the pressure is

$$< \rho_s(1) \frac{\partial w_s^{es}(1)}{\partial z_1} >_S = -\frac{4\pi\mu_s}{3} \int_0^\infty 1/S \int dx_1 dy_1 \rho_s(1) \frac{\partial E_0(1)}{\partial z_1} \qquad (1.37)$$

which after a short calculation leads to

$$P = k_B T \left[\sum_{i=1}^m \bar{\rho}_i(\sigma_i/2)) + \bar{\rho}_s(\sigma_i/2)) \right] - (1/8\pi) < [E_z(1)]^2 >_S$$

$$- \sum_{i=1}^m < \rho_i(1) \frac{\partial w_i^{cov}(1)}{\partial z_1} >_S - \frac{4\pi\mu_s}{3} < \sum_m \rho_{s,m}^1 \frac{\partial E_{-m}^1(\mathbf{r}_1)}{\partial z_1} >_S \qquad (1.38)$$

where

$$\bar{\rho}_s(\sigma_s/2) = \frac{1}{(4\pi S)} \int d\phi_1 dcos\theta_1 dx_1 dy_1 \rho_s(1) \qquad (1.39)$$

and

$$\mathbf{E}_0(1) = \sum_m E_m^1(\mathbf{r}_1) \mathbf{e}_m^1 \qquad (1.40)$$

where $\mathbf{e}_m^1$ are the polar components of the unit vector.
The last term of Eq.(1.38) corresponds to a new electrostriction effect which vanishes for uniform external field $\mathbf{E}_0$.

5. The rough electrode near a non primitive electrolyte. This is a case relevant to computer simulations of realistic solvent models near a model of a metallic surface such as the silver(111) surface, for which experiments have recently been reported [61]. Most models of water employed in the computer simulations consist of neutral molecules with embedded point charges.
   The sum over the charges q_ν in each molecule is indicated by the index ν, and is zero for each molecule. Each of these charges is located at the position $\mathbf{b}_\nu$ relative to a molecular reference frame.. From Eq.(1.29) we get

$$P = k_B T \left[\sum_{i=1}^m \bar{\rho}_i(\sigma_i/2)) + \bar{\rho}_s(\sigma_i/2)) \right] - (1/8\pi) < [E_z(1)]^2 >_S$$

$$- \sum_{i=1}^m < \rho_i(1) \frac{\partial w_i^{cov}(1)}{\partial z_1} >_S + < \rho_s(1) \sum_\nu q_\nu \frac{\partial E_z(\mathbf{r}_1 + \mathbf{b}_\nu)}{\partial z_1} >_S \qquad (1.41)$$

where $\bar{\rho}_s(\sigma_i/2)$ is now defined for the orientation dependent density function of the solvent with embedded charges. Again, if there are only soft wall forces, we get

$$P = -(1/8\pi) < [E_z(1)]^2 >_S - \sum_{i=1}^{m} < \rho_i(1) \frac{\partial w_i^{cov}(1)}{\partial z_1} >_S$$

$$+ < \rho_s(1) \sum_{\nu} q_\nu \frac{\partial E_z(\mathbf{r}_1 + \mathbf{b}_\nu)}{\partial z_1} >_S \qquad (1.42)$$

This relation points to the importance of using a model potential for liquid water that has the correct equation of state (pressure) rather than the correct bulk density (off may be by a few percent) when computing density profiles near planar or rough electrodes.

1.3.2 The smooth interface

Basic definitions: Gouy-Chapman theory

We have a mixture of ions of density ρ_i, charge e_i and diameter σ_i. $\rho_i(z)$ is the number density profile of i at a distance z from the electrode, which is always assumed to be flat and perfectly smooth. The singlet distribution function is

$$g_i(z) = \frac{\rho_i(z)}{\rho_i} = h_i(z) + 1 \qquad (1.43)$$

The charge density $q(z)$ is given by

$$q(z) = \sum_{i=1}^{m} e_i \rho_i(z) \qquad (1.44)$$

where m is the number of ionic species. The electrostatic potential ϕ is obtained by integration of Poisson's equation

$$\nabla^2 \phi(z) = \frac{d^2\phi(z)}{dz^2} = \frac{-4\pi q(z)}{\epsilon} \qquad (1.45)$$

This equation can be integrated to obtain the alternate relation between the charge and potential profiles

$$\phi(z) = \frac{4\pi}{\epsilon} \int_0^\infty |z - t| q(t) dt \qquad (1.46)$$

The total potential drop $\Delta\phi$ is obtained from Eq.(1.46) by either letting $z = 0$ or $z \to \infty$, depending on the reference potential of the model. In general the latter choice is adopted. An important quantity is the differential capacitance C_d which defined by

$$C_d = \frac{dq_s}{d\Delta\phi} \tag{1.47}$$

where q_s is the surface charge on the electrode. This quantity is difficult to measure directly, and is inferred from either surface tension measurements, or frequency dependent AC measurements of the capacitance. The surface charge satisfies the electroneutrality condition

$$q_s = -\int_0^\infty q(z)dz = \frac{E_0\epsilon}{4\pi} \tag{1.48}$$

where E_0 is the external or applied field.

Consider the Poisson equation Eq.(1.45). If we approximate the density of the ions by Boltzmann' distribution formula [62, 63]

$$\rho_i(z) = \rho_i e^{-\beta e_i\phi(z)} \tag{1.49}$$

replacing into Eq.(1.45) we obtain the Poisson Boltzmann equation

$$\nabla^2\phi(z) = \frac{-4\pi}{\epsilon}\sum_{i=1}^m e_i\rho_i\, e^{-\beta e_i\phi(z)} \tag{1.50}$$

A first integral of this differential equation can be obtained multiplying both sides by $\nabla\phi(z)$. For the planar electrode this yields

$$E_0{}^2 = [\nabla\phi(z)]^2 = \frac{-8\pi}{\epsilon}\sum_{i=1}^m e_i\rho_i\{e^{-\beta e_i\phi(z)} - 1\}\nabla\phi(z) \tag{1.51}$$

Using the definition of C_d and the electroneutrality relation Eq.(1.48) we get the formula for the differential capacitance

$$C_d = \sqrt{\frac{2\pi}{\epsilon kT}}\frac{\sum_{i=1}^m e_i\rho_i[e^{-\beta e_i\phi(0)}]}{\sqrt{\sum_{i=1}^m e_i\rho_i[e^{-\beta e_i\phi(0)} - 1]}} \tag{1.52}$$

where $\phi(0) = \phi(z)|_{z=0} = \Delta\phi$ is the potential at the origin, and is equivalent to the total polarization potential of the electrode. At this point it is convenient to make a change in the variable

$$\chi(z) = e^{-\beta e\phi(z)} \tag{1.53}$$

where $e_i = z_i e$. e is the elementary charge and z_i is the electrovalence of species i. We integrate equation Eq.(1.51) to get

$$\int_{\chi(0)}^{\chi(z)} d\chi \frac{1}{\chi\sqrt{\sum_i \rho_i \chi^{z_i} - \rho_0 A}} = \sqrt{\frac{8\pi e^2}{\epsilon kT}}(z - z_0) \qquad (1.54)$$

where χ and A are integration constants. The electrovalence z_i is always a small number and the integration of the left hand side is always possible in terms of elliptic functions [64]. When $z_1 = -z_2 = 1$ the radicand of the left hand side of Eq.(1.54) is a perfect square and the integral can be performed explicitly. For the potential drop $\Delta\phi$ we obtain the implicit relation

$$\frac{\beta e E_0}{\kappa\epsilon} = 2\sinh[\Delta\phi e\beta/2] \qquad (1.55)$$

The density profile is given by

$$\rho_i(z) = \rho_i \left[\frac{[1 + z_i\alpha e^{-\kappa z}]}{[1 - z_i\alpha e^{-\kappa z}]}\right]^2 \qquad (1.56)$$

where

$$\kappa^2 = 4\pi/\epsilon kT \sum_{i=1}^{m} \rho_i e_i^2 \qquad (1.57)$$

defines the Debye screening parameter and α is given by

$$\alpha = \tanh[\Delta\phi e\beta/4] \qquad (1.58)$$

There are several remarks about the Gouy-Chapman theory: In spite of the apparent oversimplification the Poisson Boltzmann equation satisfies an overall dynamic equilibrium condition, that fixes the contact density at the electrode surface. This is the contact theorem

$$kT \sum_{i=1}^{m} \rho_i(0) = \frac{\epsilon}{8\pi}E_0^2 + kT \sum_{i=1}^{m} \rho_i \qquad (1.59)$$

This contact theorem, as well as other sum rules that are valid for the charged interface will be discussed in the next section. The density profiles obtained from the Gouy-Chapman theory are monotonous, that is they show no oscillations. Since in this theory the contact theorem and the electroneutrality condition are satisfied, then, $\rho_i(z)$ is pinned at the origin, and has a fixed integral, so that the density profile cannot deviate too much from the correct result. When the contact theorem is not satisfied, such as in the case of mixtures of unequal size ions at low electrode charge or for high density, when the profiles oscillatory, we expect deviations from the GC theory. This is also true for the non-primitive model in which the solvent is a fluid of finite size molecules.

In the regime of low density and high temperature (or large dielectric constant) the Gouy-Chapman theory is quite accurate in spite of its simplifications because it satisfies both the contact theorem Eq.(1.29) asymptotically for $E_0 \to \infty$, and the electroneutrality condition. Thus, the density profile is basically pinned at the electrode wall at the correct point, and the integral is also fixed to the correct value. However in real systems with molecular solvents the density and coupling constant are large, significant deviations from the behavior predicted by the GC theory occur, because in this case the dominant term in the contact theorem is not the charge term, but the solvent hard core term. For this reason it is interesting to assess the accuracy of the integral equations for the primitive model for high coupling constants beyond the parameters that correpond to experimental situations, because it will indicate which theory can be used for the non- primitive model of the electric double layer.

These theories can be formulated as integral equations for the density profile $\rho_i(1)$, or as a differential or integrodifferential equation for the potential $\phi(1)$, or can be derived from a functional $\Phi(1)$, which is dependent on the position. All of these theories can be derived using functional differentiation: The central quantity of our discussion is [65] the one particle direct correlation function, from which the integral equations will be deduced:

$$c_i(1) = \ln \left[\frac{\rho_i(1)}{z_i} \right] + \beta w_i(1)$$

$$= \ln \left[\rho_i(1) \right] + \beta [w_i(1) - \mu_i(1)] \tag{1.60}$$

where $c_i(1)$ is the one particle direct correlation function, z_i is the fugacity of species i, and $u_i(1)$ is the external potential. Furthermore

$$z_i = e^{\beta \mu_i(1)} \tag{1.61}$$

The function $c_i(1)$ is a member of the family of direct correlation functions $c_{ij..}(1, 2, ..)$, which is the sum of all irreducible graphs with density factors $\rho_i(1)$ for every field point (For a detailed discussion of correlation functions see for example Hansen and McDonald [66]). The integral equations can be obtained by differentiation of this magnitude. Functional series differentiation[67, 68] produces approximations, such as the Hypernetted Chain (HNC) and its modifications, and the Mean Spherical (MSA) and its modifications that are used in conjunction with the Ornstein Zernike equation. A different set of approximations is obtained by spatial differentiation of $c_i(1)$, which gives the Born Green Yvon (BGY) and Wertheim Lovett Mou Buff equations (WLMB). Finally the Kirkwood equation is obtained by differentiation with respect to the chemical potential.

The BGY equation [69, 70] can be derived from the one particle direct correlation function $c_i(1)$. Consider again Eq.(1.71): letting the gradient ∇ act on the f of the graphical expansion of $c_i(1)$, we get the BGY equation: The first member of this hierarchy is:

$$-kT \,\nabla_1 \,\rho_i(1) = \rho_i(1) \,\nabla_1 \,u_i(1) + \sum_{j=1}^{m} \int d2\rho_{ij}(1,2) \,\nabla_1 \,u_{ij}(1,2) \qquad (1.62)$$

Using Eq.(1.74) and Eq.(1.77) to eliminate the long ranged terms, we obtain Eq.(1.62) in a different form

$$-kT \,\nabla_1 \,\rho_i(1) = \rho_i(1) \,\nabla_1 \,u_i^0(1) + e_i\rho_i(1) \,\nabla_1 \,\phi(1)$$

$$+ \sum_{j=1}^{m} \int d2\rho_{ij}(1,2) \,\nabla_1 \left[u_{ij}^0(1,2)\right] \rho_i(1)e_i \sum_{j=1}^{m} e_j \int d2\rho_j(2)h_{ij}(1,2) \,\nabla_1 \left[\frac{1}{\epsilon r_{ij}}\right] \quad (1.63)$$

This equation can be integrated from z to ∞, to yield

$$\ln[g_i(z)] = -e_i[\phi(z) + \psi_i(z)] + J_i(z) \qquad (1.64)$$

which together with Poisson equation Eq.(1.45) forms a closed system of equations that is very convenient for numerical solutions. The right hand side term consists of three contributions: The potential $\phi(1)$, which is determined by the single particle distribution function $\rho_i(z)$, and the terms $\psi_i(z)$ and J_i which are functions of the pair distribution function $h_{ij}(1,2)$. From Eq.(1.63) we get

$$\psi_i(z) = \int_z^{\infty} dz_1\rho_i(1)e_i \sum_{j=1}^{m} e_j \int d2\rho_j(2)h_{ij}(1,2) \,\nabla_1 \frac{1}{\epsilon r_{ij}} \qquad (1.65)$$

$$J_i(z) = \int_z^{\infty} dz_1 \sum_{j=1}^{m} \int d2\rho_{ij}(1,2) \,\nabla_1 \,u_{ij}^0(1,2) \qquad (1.66)$$

We remark that in Eq.(1.63) (and also in Eq.(1.64), if the fluctuation terms $J_i(z)$ and $\psi_i(z)$ are neglected, then we get back the Gouy Chapman equation Eq.(1.50), which has a known analytical solution. In the BGY based theories the pair correlation function $h_{ij}(1,2)$ must be given by some approximation. The interesting

feature of the BGY equation is that for no matter which closure, the contact theorem Eq.(1.29) is satisfied. Different approximations for the inhomogeneous pair
correlation functions $h_{ij}(1,2)$ have been studied: Torrie and Valleau [71, 72] have
made the comparison to the computer simulations for a 1-1 salt near a flat electrode
with surface charge $\sigma^* = q_s\sigma^2/e = 0.7$ [73, 74].

The comparison to the Monte Carlo simulations is good, this method yields for
the test case with the observed density oscillations in the profile of the counterions.
Another integral equation is derived from the one particle direct correlation function $c_i(1)$ Eq.(1.72) by introducing relative coordinates for all field in the diagram
representation and taking the derivatives with respect to those coordinates. This
yields and exact hierarchy of equations that is related to thy BGY hierarchy. The
first member of the Wertheim-Lovett-Mou-Buff (WLMB) equation is

$$\nabla_1\rho_i(1) + \beta\rho_i(1)\nabla_1 u_i(1) = \rho_i(1)\sum_{j=1}^{m}\int d2c_{ij}(1,2)\nabla_2\rho_j(2) \qquad (1.67)$$

This equation contains long range, divergent terms. Introducing the local potential
$\phi(1)$ Eq.(1.46), we have

$$\nabla_1 \ln\rho_i(1) + \beta\nabla_1\phi_i(1) = \sum_{j=1}^{m}\int d2c_{ij}^{sr}(1,2)\nabla_2\rho_j(2) \qquad (1.68)$$

This equation has been studied by Henderson and Plischke [75, 76, 77, 78] in detail.
It yields very good results for the test case of $\sigma^* = q_s\sigma^2/e = 0.7$.

The calculations were performed solving both the HNC2 closure for the inhomogeneous pair correlation function, and also the MSA2 closure in a few cases. A
simplified version of the WLMB equation that produces reasonably good results was
studied by Colmenares and Olivares [77, 78].

Hypernetted chain equations

At the interface between an electrode and a fluid the density of the fluid is
a function of the distance of the point to the surface $\rho_i(z)$. The Ornstein Zernike
equation for this system can be obtained as a limit of a system that is a homogeneous
mixture in which there are some large ions, of radius $R_w \to \infty$,such that $\rho_w R_w^3 \to 0$.
In this limit the planar [79, 80] HAB (Henderson-Abraham-Barker) OZ equation is

$$h_i(1) - c_i^w(1) = \sum_{j=1}^{m}\int d2h_j(2)\rho_j c_{ij}^B(1,2) \qquad (1.69)$$

where $h_i(1)$ is the density profile correlation function of ion i, $c_i^w(1)$ is not the single particle direct correlation function $c_i(1)$, but a different magnitude defined below Eq.(1.71), and $c_{ij}^B(1,2)$ is the bulk direct correlation function.

$$h_i(1) = g_i(1) - 1 = \frac{\rho_i(1) - \rho_i}{\rho_i} \tag{1.70}$$

The function $c_{ijk..}^B(1,2,3..)$ is a much more complicated object, and in general does not admit a simple diagram expansion. The understanding of the meaning of this function is clarified using a functional series expansion: Consider the functional power series expansion of $\ln \rho_i(1)$ around the uniform density [67, 68] ρ_i

$$\beta u_i(1) + \ln \rho_i(1) = \ln \rho_i + \sum_{j=1}^{m} \int d2 h_j(2) \rho_j c_{ij}^B(1,2)$$

$$+ \sum_n 1/n! \sum_{j,k..=1}^{m} \rho_j \rho_k.. \int d2 d3..h_j(2) h_k(3) c_{ijk..}^B(1,2,3..) \tag{1.71}$$

the direct correlation functions are defined by the functional derivative

$$c_{ijk..}^B(1,2,3..) = \frac{\delta^n c_i^B(1)}{\delta \rho_j(2) \delta \rho_k(3)...} \tag{1.72}$$

The superscript B stands for the bulk functions. We now introduce the function $c_i^w(1)$, defined by

$$c_i^w(1) = -\beta u_i(1) - \ln g_i(1) + h_i(1)$$

$$+ \sum_{j,k=1}^{m} \left[\rho_j \rho_k.. \int d2 d3 h_j(2) h_k(3) c_{ijk}^B(1,2,3) \right] + ... \tag{1.73}$$

The inhomogeneous potential is of the form

$$u_i(1) = u_i^0(1) + w_i(1) \tag{1.74}$$

with $u_i^0(1)$ the short ranged and for a hard, smooth charged electrode, the electrostatic part is

$$w_i(1) = -e_i E_0 z_1/2 \tag{1.75}$$

Combining this definition with the functional expansion Eq.(1.71) we get the HNC1 equation for the flat wall electrode

$$-\beta w_i(1) - \ln \rho_i(1) = \sum_{j=1}^{m} \rho_j \int d2 h_j(2) c_{ij}^B(1,2) \tag{1.76}$$

Equation Eq.(1.76) has a deceivingly simple aspect, but because of the long range character of $w_i(1)$ is not convergent, and therefore not amenable to numerical solution. We write now

$$c_{ij}(|r_{12}|) = c_{ij}^0(|r_{12}|) - w_{ij}(|r_{12}|)$$ (1.77)

with

$$w_{ij}(|r_{12}|) = \frac{e_i e_j}{\epsilon |r_{12}|}$$ (1.78)

and replacing into Eq.(1.76) yields

$$-\beta e_i \phi(1) - \ln \rho_i(1) = \sum_{j=1}^m \rho_j \int d2 h_j(2) c_{ij}^0(1,2)$$ (1.79)

where $\phi(1)$ is defined by

$$\phi(1) = E_0 z_1 + \int d2 \sum_{j=1}^m \frac{e_j \rho_j(2)}{\epsilon r_{12}}$$ (1.80)

This equation is the plane electrode version of the Hypernetted Chain equation, called the HNC1 [81]. It is completely defined in terms of short ranged quantities, which is not the case for the first form of the equation Eq.(1.76). The HNC1 is the theory that has the closure with the largest number of graphs. It satisfies the electroneutrality relations and the Stillinger Lovett sum rules. One important observation about the HNC1 is that it does not satisfy the contact theorem Eq.(1.29), but rather

$$kT \sum_{i=1}^m \rho_i(0) = \frac{\epsilon}{8\pi} E_0{}^2 + \rho_0 \partial P / \partial \rho_0$$ (1.81)

where

$$\rho_0 = \sum_{i=1}^m \rho_i$$ (1.82)

For high fields and low concentrations the fact that we get the compressibility rather than the pressure is not very important and the HNC1 is still a reasonably good theory, as will be shown below. However for dense systems this is a rather severe shortcoming. Specifically, when we are dealing with a molecular (dipolar) solvent the density is very large and the dielectric constant ϵ is of the order of one (instead of 80 in water) which makes the electrostatic term in satisfies the contact theorem Eq.(1.29) small in comparison to the contact density term. The consequence is that the HNC1 will put more counter ions near the electrode than the exclusion of the hard cores will permit. Eventually thermodynamic stability conditions will be violated, and we get a negative capacitance, reflected by a decreasing potential drop $\Delta \phi$ with increasing applied external field E_0.

The HNC is the most accurate theory for bulk electrolytes. One would expect that this fact would remain true in the plane electrode limit. However, because of the inaccuracy of the HNC for uncharged hard sphere fluids the HNC1 does no do well in representing the exclusion volume of the ions, and is not on the whole, such a good approximation for the electric double layer. The bulk direct correlation function

$$c_{ij}^B(|r_{12}|) \tag{1.83}$$

which should be used in solving the HNC1 equation Eq.(1.79) is that obtained of the bulk HNC equation for the same system. This however, sometimes called the HNC/HNC approximation, yields poor results when compared to computer simulations[81]. Generally better results are obtained if instead of the HNC bulk direct correlation function the corresponding MSA functions are used, the general a-greement with computer simulations improves [82, 83, 84, 85, 86, 87]. The next term to be considered is the third term of Eq.(1.71) which is a three particle contribution. The three particle direct correlation function is in general a very complex function, and must be approximated. The simplest of these approximations is to include the first diagram of the density expansion of the three point direct correlation function , the bridge diagram [89]. Ballone, Pastore and Tosi[88] performed this calculation with good success . The density profile for the $1M, 1-1$ electrolyte at a surface charge $\sigma^* = q_s\sigma^2/e = 0.7$, which will be the test case used for comparisons. This is the highest surface density simulated, and shows charge oscillations due to the hard core of the electrolyte. In this calculation the bridge diagrams were computed directly from the product of the three bulk pair correlation functions, which is first term in the density expansion of the bulk triplet direct correlation function

$$c_{ijk}^B(1,2,3) = h_{ij}^B(1,2)h_{ik}^B(1,3)h_{kj}^B(3,2) \tag{1.84}$$

Since there are no adjustable parameters, the agreement is very good. An alternative less laborious procedure was suggested by Rosenfeld and Blum [89], but actual calculations were not performed. Another way of inmproving the HNC1 approximation was introduced by Forstmann and co-workers [90, 91, 92, 93, 94]. In their method the HNC1 equation is used as described above, but instead of taking the bulk direct correlation function, as prescribed by Eq.(1.79), a local density dependent $c_{ij}^B(r, \bar{\rho})$ is taken. The local density is defined by

$$\bar{\rho}_i(z) = \frac{1}{2\sigma\delta} \int_{z-\delta}^{z+\delta} dx \int_{x-\sigma/2}^{x+\sigma/2} dy \rho_i(y) \tag{1.85}$$

where σ is the diameter of the ion and δ is an adjustable parameter. The bulk correlation function is then

$$c_{ij}^B(|r_{12}|)|_{\rho_i=\bar{\rho}_i(z)} \tag{1.86}$$

For the test case with surface charge $\sigma^* = .7$, the results of this method are very good.

Kirkwoods equation

An interesting approach has been suggested by Kjellander and Marcelja [95, 96, 97], based on the observation that for the HNC approximation the chemical potential can be obtained explicitly as a function of the pair potential $h_{ij}(|r_{12}|)$ for an homogeneous fluid. Then, within the HNC the function $c_i(1), c(1)$ can be explicitly evaluated. The central idea is to slice the three dimensional space into two dimensional layers that are homogeneous. The three dimensional OZ equation can be mapped into coupled set of N two dimensional OZ equations for a mixture of N components, each component is an ion in a different layer. The particles interact with a species dependent interaction pair potential. In the limit of an infinite number of layers this procedure yields the correct inhomogeneous OZ equation. The chemical potential $\mu_i(\alpha)$ of the ith ion in the αth layer is given by Kirkwoods equation:

$$\mu_i(\alpha) = kT \ln \rho_i(\alpha) + kT \ln \frac{\Lambda_0}{\Delta z}$$

$$+ V_i(\alpha) + \sum_{j=1}^{m} \rho_j(\beta) \int_0^1 d\lambda \int dR g_{ij}(R, \alpha\beta; \lambda)$$

$$\frac{\partial[g_{ij}(R, \alpha\beta; \lambda)]}{\partial \lambda} \tag{1.87}$$

where λ is the coupling parameter, Δz is the thickness of the layer, Λ_0 is the ideal gas fugacity, $V_i(\alpha)$ is the interaction between a particle in layer α and the wall; R is the two dimensional distance. In the HNC closure

$$c_{ij}(R, \alpha\beta) = -\beta w_{ij}(R, \alpha\beta) + h_{ij}(R, \alpha\beta) - \ln g_{ij}(R, \alpha\beta) \tag{1.88}$$

Kirkwoods equation can be integrated to yield

$$\rho_i(\alpha) = \frac{\Delta z}{\Lambda_0}$$

$$exp[\beta\mu_i(\alpha) + \sum_{\beta,j=1} \rho_j(\beta) \int dR[(1/2)h_{ij}^2(R, \alpha\beta) - c_{ij}(R, \alpha\beta) - \beta w_{ij}(R, \alpha\beta)]$$

$$- [(1/2)\ln[g_{ij}(R, \alpha\beta)] - \beta w_{ij}(R, \alpha\beta)/2]_{R=0} - \Phi_i(\alpha)]_{inh} \tag{1.89}$$

where $\Phi_i(\alpha)$ is the average potential for layer α.

$$\Phi_i(\alpha) = \frac{2\pi e^2}{\epsilon} \sum_{\beta} \rho_i(\beta)|z_\alpha - z_\beta| \tag{1.90}$$

Density functionals

The density functional method has proven to be one of the more succesful and versatile technique to study interfacial and bulk phenomena. The central quantity in the density functional is the excess (over ideal gas) free energy which originates from the interactions amongst the particles. In classical mechanics it is a uniquely defined functional of the spatially varying one particle density $\rho(\mathbf{r})$, from which many equilibrium properties of the fluid can be derived. The most succesful density functionals are the non local free energy density functionals, which employ weighted or locally averaged densities that are constructed to fit available structural and thermodynamic properties of the homogeneous fluid.

A remarkable functional expansion for the inhomogeneous fluids was developed by Rosenfeld [98, 99, 100, 101].

In this approach liquid state theories like the MSA and the HNC can be derived as variational problems of the free energy functional, which is written in terms of the Ornstein-Zernike direct correlation functions of order 1,2, Eq.(1.60) . [98, 99, 100, 101] These correlation functions can be expressed in terms of a reduced set of basis functions which are related to the geometry of the molecules in the fluid. In the asymptotic limit of strong Coulomb interactions between the charged particles , that is the limit in which either the charge goes to infinity or the temperature goes to zero, [102] the free energy and the internal diverge to the same order in the the the coupling parameter that is

$$\frac{\Delta E}{\Delta F} \to 1 \tag{1.91}$$

while the entropy diverges at a slower rate. In this asymptotic limit, the free energy and the energy coincide, and furthermore, the mean spherical approximation (MSA) and the hypernetted chain approximation (HNC) concide. This is a very gratifying feature, because the HNC, which from the diagram expansion (and numerous test cases) point of view is the more accurate theory, is in general difficult to solve , while the MSA is analytical in most cases, and in the asymptotic limit, of a rather surprisingly simple form. In the asymptotic limit the excess electrostatic energy is identical to the exact Onsager lower bound, which is achieved by immersing the entire hard core system in an infinite neutral and perfectly conducting (liquid metal) fluid. The Onsager process of introducing the infinite conductor, naturally decouples all the differeint components in the system which may differ in size, shape, charge distribution an relative oreintation in space.

The direct correlation function in the asymptotic strong coupling limit (Onsager picture) is obtained directly from the electrostatic interaction of the charges of the particles smeared on the surface of those particles.

Another asymptotic limitis the <u>high density limit</u>, in which the compressibility tends to zero because of the tight packing of the particles. In this case the MSA solution is also obtained from a simple geometric argument by computing the overlap volume of the particles as a function of their distance and their relative orientation. These two distinct limits provide the set of basis functions for the representation of the direct correlation function, which can be shown to be sufficient to represent the dcf of the complete MSA solution. In other words, these two limits provide the full <u>functional basis set</u> for the exact solution of the MSA equations and also an asymptotic approximation of the HNC solution for all densities and temperatures, for hard charged objects. The basis functions for the functional expansion of the direct correlation function are obtained from linear combinations of overlap functions, such as the volume,the surface and the convex radius, and the electrostatic interaction between surface smeared charges. By proper manipulation of the free parameters, and by a judicious selection of the basis set of trial functions, one can obtain, different levels of approximations. The physically intuitive meaning of the basis functions in the representation of the dcf is particularly illuminating in the formulation of perturbation treatments. The use of the asymptotic basis set of functions ensures that at all levels of the perturbation approximation, the resulting free energy has the desired property of interpolating between two exact lower bounds, the Debye Hueckel result (which is effective at weak coupling) and the Onsager result, (which is effective at high coupling). These two limits pin the free energy.

The interpolation between the low and high density limits, which is inherent to this variational approach, leads in a very natural way to the scaled particle theory for the structure and thermodynamics of isotropic fluids of hard particles. This unifies, for the first time the Percus Yevick theory, which is based on diagram expansions, and the scaled particle theory of Reiss, Frisch and Lebowitz, and, at the same time yields the analytical expressions of the dcf conformal to those of the hard spheres. It provides an unified derivation of the most comprehensive analytic description available of the hard sphere thermodynamics and pair distribution functions as given by the Percus Yevick and scaled particle theories, and yields simple explicit expressions for the higher direct direct correlation functions of the uniform fluid.

Hard sphere fluids

For the inhomogeneous fluid of hard spheres characterized by the set of of one particle densities $\rho_i(\mathbf{r})$ the free energy functional is

$$F = F^{id} + F^{ex} \tag{1.92}$$

where

$$F^{id} = \sum_i \int d\mathbf{r} \rho_i(\mathbf{r}) \left[\lambda_i^3 \log \rho_i(\mathbf{r}) - 1 \right] \tag{1.93}$$

$$F^{ex} = \int d\mathbf{r}\Phi[\{n_\alpha(\mathbf{r})\}] \tag{1.94}$$

where $\Phi[\{n_\alpha(\mathbf{r})\}]$ is the excess free energy density, which is a function of the system's averaged geometric measures

$$n_\alpha(\mathbf{x}) = \sum_i \int d\mathbf{r}\rho_i(\mathbf{r})\omega_i^{(\alpha)}(\mathbf{r} - \mathbf{x}) \tag{1.95}$$

The functional Φ that reproduces the scaled particle theory and the Percus-Yevick copressibility equation of state for uniform hard sphere liquids is

$$\Phi = -n_0 \log(1 - n_3) + \frac{n_1 n_2}{(1 - n_3)} + \frac{n_2^3}{24\pi(1 - n_3)^2} \tag{1.96}$$

where

$$n_\alpha = \sum_i \rho_i R_i^{(\alpha)} \tag{1.97}$$

or

$$n_3 = (4\pi/3) \sum_i \rho_i R_i^3 \qquad n_2 = (4\pi) \sum_i \rho_i R_i^2 \sum_i \frac{\partial n_3}{\partial R_i} \tag{1.98}$$

$$n_1 = (1/8\pi) \sum_i \frac{\partial n_2}{\partial R_i} = \sum_i \rho_i R_i \qquad n_0 = \sum_i \rho_i \sum_i \frac{\partial n_1}{\partial R_i} \tag{1.99}$$

From Eq.(1.95), we deduce

$$\frac{\delta n_\alpha}{\delta \rho_i(\mathbf{r}_1)} = \omega_i^{(\alpha)}(\mathbf{r}_1) \tag{1.100}$$

so that one possible set of weight functions is

$$\omega_i^{(3)}(\mathbf{r}) = \theta(R_i - |r|)$$
$$\omega_i^{(2)}(\mathbf{r}) = \delta(R_i - |r|)$$
$$\omega_i^{(1)}(\mathbf{r}) = (1/8\pi)\delta'(R_i - |r|)$$
$$\omega_i^{(0)}(\mathbf{r}) = -(1/8\pi)\delta''(R_i - |r|) - (1/2\pi r)[\delta'(R_i - |r|) \tag{1.101}$$

The thermodynamic properties of hard sphere mixtures are expressed in terms of the basis n_α : For example it is verified that the PY compressibility pressure is

$$\beta P = \frac{\partial \Phi}{\partial n_3} = -\frac{n_0}{(1 - n_3)} + \frac{n_1 n_2}{(1 - n_3)^2} + \frac{n_2^3}{12\pi(1 - n_3)^3} \tag{1.102}$$

In the uniform fluid the free energy density obeys the scaled particle differential equation

$$\beta P = \sum_\alpha \frac{\partial \Phi}{\partial n_\alpha} + n_0 - \Phi \tag{1.103}$$

which can be verified directly.

The chemical potential can also be expressed in terms of Φ:

$$\frac{\delta F}{\delta \rho_i(\mathbf{r}1)} = c_i(1) = \ln[\rho_i(1)] + \beta[w_i(1) - \mu_i(1)] \tag{1.104}$$

Else

$$\frac{\delta \Phi}{\delta \rho_i(\mathbf{r}1)} = -\beta \mu_i(1) \tag{1.105}$$

$$= \sum_\alpha \frac{\partial \Phi}{\partial n_\alpha} \frac{\delta n_\alpha}{\delta \rho_i(\mathbf{r}1)} \tag{1.106}$$

or, in other words

$$-\beta \mu_i(1) = \sum_\alpha \frac{\partial \Phi}{\partial n_\alpha} w_i^{(\alpha)}(\mathbf{r}_1) \tag{1.107}$$

The charged case

Consider the electrostatic charge part of the excess free energy $F^{es}[\{\rho_i(\mathbf{r})\}]$ and use the functional expansion formalism to expand around the bulk density $[\{\rho_i\}]$ The variable in this case is

$$\Delta \rho_i(\mathbf{r}) = \rho_i(\mathbf{r}) - \rho_i$$

We get, up to second order,

$$F^{es}[\{\rho_i(\mathbf{r})\}] = F^{es}[\{\rho_i\}] - (1/\beta) \sum_i c_i^{(1),es}[\{\rho_i\}] \int d\mathbf{r} \Delta \rho_i(\mathbf{r})$$

$$-(1/2\beta) \sum_i \int d\mathbf{r} d\mathbf{r}' c_{i,j}^{(2),es}[\{\rho_i\}](|\mathbf{r} - \mathbf{r}'|) \Delta \rho_i(\mathbf{r}) \Delta \rho_i(\mathbf{r}') \tag{1.108}$$

Here we have used the defining relations

$$c_i^{(1),es}[\{\rho_i\}(\mathbf{r}_1)] = -\beta \frac{\delta F^{es}}{\delta \rho_i(\mathbf{r}_1)} \tag{1.109}$$

and

$$c_{i,j}^{(2),es}[\{\rho_i\}](\mathbf{r}_1, \mathbf{r}_2) == -\beta \frac{\delta^2 F^{es}}{\delta \rho_i(\mathbf{r}_1)} \delta \rho_j(\mathbf{r}_2) \tag{1.110}$$

The one point electrostatic direct correlation function is minus the excess chemical potential,

$$c_i^{(1),es}[\{\rho_i\}] = -\beta \mu_i[\{\rho_i\}] \tag{1.111}$$

while the two point direct correlation function in Eq.(1.108)

$$c_{i,j}^{(2),es}[\{\rho_i\}](|\mathbf{r} - \mathbf{r}'|)$$

is taken at the uniform densities of the bulk fluid.

Using the complete free energy functional Eq.(1.92) for the hard core part and the truncated expansion for the charge part, we obtain the functional

$$F^{ex}(\mathbf{r}) = F^{hc}(\mathbf{r}) + F^{es}(\mathbf{r} = \infty) - (1/\beta) \sum_i c_i^{(1),es}[\{\rho_i\}] \int d\mathbf{r} \Delta \rho_i(\mathbf{r})$$

$$-(1/2\beta) \sum_i \int d\mathbf{r} d\mathbf{r}' c_{i,j}^{(2),es}[\{\rho_i\}](|\mathbf{r} - \mathbf{r}'|) \Delta \rho_i(\mathbf{r}) \Delta \rho_j(\mathbf{r}') \tag{1.112}$$

An alternative derivation was given by Kierlik and Rosinberg [103], following essentially that of Sluckin and Evans [104]

$$F(\mathbf{r}) = F^{hc}(\mathbf{r}) + (1/2) \int d\mathbf{r} q \Psi(\mathbf{r}) + \sum_i \int d\mathbf{r} \rho_i(\mathbf{r})[w_i^{cov}(\mathbf{r}) - \mu_i] \tag{1.113}$$

where $F^{hc}(\mathbf{r})$ is the full functional for the hard core interaction $w_i^{cov}(\mathbf{r})$ is the non electrostatic part of the external potential; Furthermore

$$q(\mathbf{r}) = q_s + \sum_{i=1}^{m} ez_i \rho_i(\mathbf{r}) \tag{1.114}$$

is the total charge density. The potential $\Psi(\mathbf{r})$

$$\Psi(\mathbf{r}) = \frac{1}{\epsilon} \int d(\mathbf{r}') \frac{q(\mathbf{r}')}{(|\mathbf{r} - \mathbf{r}'|)} \tag{1.115}$$

The functional inverse of this equation is Poisson's equation

$$\nabla^2 \Psi(\mathbf{r}) = -\frac{4\pi}{\epsilon} q(\mathbf{r}') \tag{1.116}$$

Requiring that

$$\frac{\delta F}{\delta \rho_i(\mathbf{r})} = 0$$

implies the Euler-Lagrange equations

$$\frac{\delta F^{hc}}{\delta \rho_i(\mathbf{r})} + ez_i \Psi(\mathbf{r}) + w_i^{cov}(\mathbf{r}) = \mu_i \tag{1.117}$$

The variational requirement

$$\frac{\delta F}{\delta \Psi(\mathbf{r})} = 0$$

which from Eq.(1.113) is equal to

$$q(\mathbf{r}) + \int d(\mathbf{r}') \frac{\delta q(\mathbf{r}')}{\delta \Psi(\mathbf{r})} \Psi(\mathbf{r}') = 0 \tag{1.118}$$

but we have the relation of the functional derivatives and their inverse

$$\int d(\mathbf{r}') \frac{\delta q(\mathbf{r})}{\delta \Psi(\mathbf{r}')} \frac{\delta \Psi(\mathbf{r}')}{\delta q(\mathbf{r}")} = \delta(\mathbf{r} - \mathbf{r}") \tag{1.119}$$

but from Eq.(1.115) and Eq.(1.116) we have that

$$\frac{\delta \Psi(\mathbf{r}')}{\delta q(\mathbf{r}")} = \frac{1}{\epsilon} \frac{1}{(|\mathbf{r}" - \mathbf{r}'|)} \tag{1.120}$$

and

$$\frac{\delta q(\mathbf{r})}{\delta \Psi(\mathbf{r}')} = -\frac{4\pi}{\epsilon} \delta"(\mathbf{r} - \mathbf{r}") \tag{1.121}$$

from where we can show that, indeed, Eq.(1.118) is equivalent to Eq.(1.116).

$$F^{ex}[\{\rho_i\}] = \int d\mathbf{r} g^{ex}[\{\rho_i\}] + \tilde{\mu}_i^{ex} \int d\mathbf{r} \Delta \rho_i(\mathbf{r})$$

$$-(1/2\beta) \sum_j \int d\mathbf{r} d\mathbf{r}' c_{i,j}^{sr}[\{\rho_i\}](|\mathbf{r} - \mathbf{r}'|) \Delta \rho_i(\mathbf{r}) \Delta \rho_j(\mathbf{r}') \tag{1.122}$$

where we have used the relationships between the functional derivatives of F^{ex} and the non Coulomb parts of the direct correlation functions. Furthermore,

$$\tilde{\mu}_i^{ex} = \mu_i^{ex} - ez_i \Psi(bulk) \tag{1.123}$$

is the contribution from the excess chemical potential arising from the non Coulomb terms, and g^{ex} is the Helmholtz excess free energy density of the uniform ionic mixture minus the electrostatic self energy. We write

$$c_{i,j}^{sr}[\{\rho_i\}] = c_{i,j}^{hc}[\{\rho_i\}] + c_{i,j}^{es}[\{\rho_i\}] \tag{1.124}$$

where c^{hc} is the direct correlation function of the uniform hard spheres fluid. Up to quadratic terms, then

$$F[\{\rho_i\}] = F^{hc}[\{\rho_i\}] + \int d\mathbf{r}[g^{ex}[\{\rho_i\}] - f_i^{hc}[\{\rho_i\}] + [-\tilde{\mu}_i^{hc}] \int d\mathbf{r}\Delta\rho_i(\mathbf{r})$$

$$-(1/2\beta)\sum_j \int d\mathbf{r}d\mathbf{r}'c_{i,j}^{(2),es}[\{\rho_i\}](|\mathbf{r} - \mathbf{r}'|)\Delta\rho_i(\mathbf{r})\Delta\rho_j(\mathbf{r}') \tag{1.125}$$

where μ_i^{hc} and f_i^{hc} are the chemical potentials and the Helmholtz free energy of the uniform hard sphere fluid. From this equation we get the Euler-Lagrange equation for the equilibrium density profiles

$$\ln[\rho_i(\mathbf{r})/\rho_i] + \beta w_i(\mathbf{r}) = \left[\frac{\delta F^{hc}}{\delta\rho_i(\mathbf{r})} - \mu_i^{hc}\right]$$

$$+ez_i[\Psi(\mathbf{r}) - \Psi(bulk)] - (1/\beta)\sum_j \int d\mathbf{r}'c_{i,j}^{(2),es}[\{\rho_i\}](|\mathbf{r} - \mathbf{r}'|)\Delta\rho_j(\mathbf{r}') \tag{1.126}$$

This equation can be obtained by integration of the WLMB equation, and in fact, it can be shown that the Rosenfeld theory of the MSA bulk is exactly equivalent to the WLMB.

The results of the Rosenfeld density functional theory are shown in figure 1.4.

As can be seen in the figure 1.4, it yields very good results for the test case of $\sigma^* = q_s\sigma^2/e = 0.7$.

MSA for ion-dipole mixtures

A very simple picture of the structure and fields near the electrode surface can be obtained from the MSA [105, 106, 107, 108, 109]. The solution of the MSA is completely explicit for very dilute solutions. We get for the charged interface

$$g_i(x) = g^{hs}(x) - \frac{\beta z_i eE}{\epsilon\kappa}e^{-\kappa(x-\sigma_i:)} \tag{1.127}$$

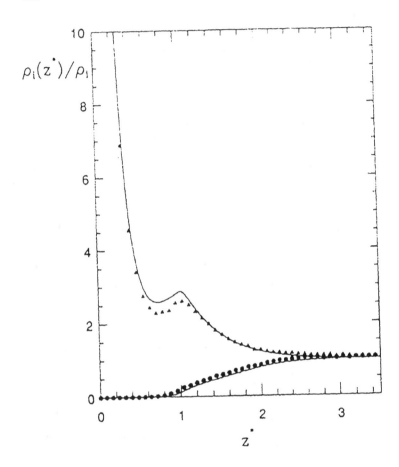

Figure 1.4: Ionic density profiles for 1:1 electrolyte (restricted primitive model) near a charged hard wall. Bulk concentration of 1 mol L^{-1} and $\sigma^* = 0.7$. The theoretical results are shown by solid line, while simulation results are shown as circles (co-ions) and triangles (counter ions).

where σ_i is the hard ion diameter and $g^{hs}(x)$ is the hard spheres against a hard wall profile [109] The profile of the dipoles is

$$g_s(x, \theta) = g^{hs}(x) - \sqrt{3}\Delta h_s(x) \cos \theta \qquad (1.128)$$

The potential difference across the interface is

$$V = 4\pi e \sum_i z_i \rho_i \int_0^\infty t h_i(t) dt + 4\pi/\sqrt{3}\rho_s \mu_s \int_0^\infty h_s(t) dt = I_1 + I_2 \qquad (1.129)$$

where ρ_i and ρ_s are the densities of the ions of species i and solvent s; μ_s is the solvent dipole moment. The first integral yields

$$I_1 = E(1 + \kappa_0 \sigma/2)/\kappa_0 \qquad (1.130)$$

while the second yields

$$I_2 = E(1 - 1/\epsilon)[1 + \kappa_0 \sigma/2(1 - 1/\lambda)]/\kappa_0 \qquad (1.131)$$

The parameter λ is calculated from the dielectric constant ϵ using Wertheim's formula

$$\lambda^2(\lambda + 1)^4 = 16\epsilon \qquad (1.132)$$

from where we get the MSA result for the potential drop

$$V = \frac{E}{\epsilon\kappa} + \frac{E\sigma}{2\epsilon}\left(1 + \frac{\epsilon - 1}{\lambda}\right) + ... \qquad (1.133)$$

It is to be noted that there has been a cancellation between terms of order 1 and $(1 - \epsilon)/\epsilon$ to produce a term of order $1/\epsilon$. The capacitance for the linearized Guoy-Chapman theory then yields

$$1/C = dV/dE = V/E = 4\pi\left[\epsilon\kappa_0 + (\sigma/2\epsilon)\left(1 + \frac{\epsilon - 1}{\lambda}\right)\right] \qquad (1.134)$$

The MSA predicts that the capacitance is equivalent to that of two capacitors in series, and that C_H, the solvent term

$$1/C_H = 2\pi(\sigma/\epsilon)\left(1 + \frac{\epsilon - 1}{\lambda}\right) \qquad (1.135)$$

is independent of concentration. This result is true also for systems with different sizes ions and dipoles. A full nonlinear extension of this treatment is available from Patey and Torrie.

A nonlinear treatment using the HNC approximation is an attractive possibility. Recently, Patey et al.[111] have solved this equation for the bulk electrolyte.

1.3.3   Adsorption at structured interfaces

A simple model of chemisorption on structured interfaces, such as crystals or biological molecules ,is one in which the binding process of individual atoms is described by an adsorption site with a binding free energy or affinity, as first proposed by Langmuir. This will neglect the details of the adsorption process, but will generate a model with a complete description of the cooperative effects that take place at the surface[56, 57, 55] via a mean field. This model, the sticky site model (SSM), is a combination of the sticky potential model, first proposed by Boltzmann, but developed by Baxter[112], and the adsorption site model of Langmuir. It is a natural way to embed the smooth interface discussed in last section in the discrete structure of the adsorbate.

Consider the case of sticky spheres: The sticky potential has the form

$$e^{-\beta u(\mathbf{r})} = 1 + \lambda \delta(r - \sigma^-), \tag{1.136}$$

where $\beta = 1/kT$ is the usual Boltzmann thermal factor, $u(\mathbf{r})$ is the intermolecular potential, λ is the stickiness parameter, $\mathbf{r} = (x, y, z)$ is the relative position of the center of the molecules, and σ is the diameter of the molecules. The right hand side term represents the probability of two molecules being stuck by the potential $u(\mathbf{r})$: this occurs only when the two molecules are in contact, and for this reason we use the Dirac delta function $\delta(r - \sigma^-)$, which is zero when the molecules do not touch, is infinity when they do, but the integral is normalized to one. The parameter λ represents the likelyhood of adsorption of an individual molecule onto the surface.

The Langmuir adsorption sites can be represented by a collection of sticky sites of the same form as was suggested by Baxter. Only that now we do not have a sphere covered uniformly by a layer of glue, but rather a smooth, hard surface with sticky points, which represent adsorption sites where actual chemical bonding takes place. For a regular crystal lattice face, Eq.(1.136) has to be changed to

$$e^{-\beta U^s(\mathbf{r})} = 1 + \lambda(\mathbf{R})\delta(z), \tag{1.137}$$

with

$$\lambda(\mathbf{R}) = \sum_{n_1, n_2} \lambda \delta(\mathbf{R} - n_1 \mathbf{a_1} - n_2 \mathbf{a_2}). \tag{1.138}$$

Here z is the distance to the contact plane, which is at a distance $\sigma/2$ from the electrode, and $\mathbf{R} = (x, y)$ is the position of a point on the planar surface at z. In Eq.(1.138), n_1, n_2 are natural numbers, and $\mathbf{a_1}, \mathbf{a_2}$ are lattice vectors of a lattice $\Lambda(z)$ on the surface at z. The lattice of sticky adsorption sites at the electrode surface is

$\Lambda(-\sigma/2)$. The requirement of point adsorption sites rather than extended regions around the sites is not essential to our discussion. It is clear that phase transitions will occur even in smooth surfaces, simply because the two dimensional gas does undergo such phase transitions. Less localized forms of the adsorption potential can be includes as long as the soft potential does not overlap neighboring sites. The model as it stands includes every interaction of the adsorbed atoms: The solvent mediated potentials of mean force as well as the quantum effects at metallic surfaces

Consider then a fluid of only one kind of particle of diameter σ, near a smooth, hard wall with sticky sites. The fluid has N particles and the volume of the system is V. The Hamiltonian of the system is

$$H = H_0 + H_S, \tag{1.139}$$

where H_0 is the Hamiltonian of the system in the absence of the sticky sites on the hard wall, and H_S is the sticky sites interaction

$$H_S = \sum_{i=1}^{N} U^s(\mathbf{r}_i), \tag{1.140}$$

where $U^s(\mathbf{r}_i)$ is the sticky interaction of Eq.(1.137). The canonical partition function of this model is

$$Z = (1/N!) \int d\mathbf{r}^N e^{-\beta H_0} \prod_{i=1}^{N} [1 + \lambda(\mathbf{R}_i)\delta(z_i)]. \tag{1.141}$$

Expanding the product in Eq.(1.141) and integrating the Dirac delta functions we get, using the single component notation to avoid heavy and unnecessarily complex equations, (with the understanding that in the multi component case N is a vector quantity with components $N_1, N_2, ..., N_n$, the necessary modifications of N! and the integrations have to be made),

$$Z = Z_0 \sum_{n=0}^{N} (\lambda^n/n!) \sum_{\{\mathbf{R}_i\} \subset \Lambda(0)} \rho_n^0(\mathbf{R}_1, \mathbf{R}_2, ..., \mathbf{R}_n), \tag{1.142}$$

where $\Lambda(0)$ is the triangular lattice at the contact plane, and where

$$\rho_n^0(\mathbf{r}_1, \mathbf{r}_2, ..., \mathbf{r}_n) = (Z_0(N-n)!)^{-1} \int d\mathbf{r}_{n+1}...d\mathbf{r}_N e^{-\beta H_0}, \tag{1.143}$$

$$= g_n^0(\mathbf{r}_1, \mathbf{r}_2, ..., \mathbf{r}_n) \prod_{i=1}^{n} \rho_1^0(\mathbf{r}_i). \tag{1.144}$$

Here $g_n^0(\mathbf{r}_1, \mathbf{r}_2, ..., \mathbf{r}_n)$ is the n-body correlation function, and $\rho_1^0(\mathbf{r}_i) = \rho_1^0(z_i)$ is the singlet density of the inhomogeneous smooth wall problem. The partition function is

$$Z_0 = (1/N!) \int d\mathbf{r}^N e^{-\beta H_0}. \tag{1.145}$$

In the sticky sites model (SSM), the excess properties of the interface depend only on the correlation functions of the smooth interface. Introducing the potentials of mean force $\omega_n(\mathbf{R}_1, \mathbf{R}_2, ..., \mathbf{R}_n)$

$$g_n^0(\mathbf{R}_1, \mathbf{R}_2, ..., \mathbf{R}_n) = e^{-\beta \omega_n(\mathbf{R}_1, \mathbf{R}_2, ..., \mathbf{R}_n)}, \tag{1.146}$$

Combining these expressions we arrive at

$$Z/Z_0 = \sum_{n=0}^{N} \left([\lambda \rho_1^0(0)]^n / n! \right) \sum_{\{\mathbf{R}_i\} \subset \Lambda(0)} e^{-\beta \omega_n(\mathbf{R}_1, \mathbf{R}_2, ..., \mathbf{R}_n)}, \tag{1.147}$$

which is the central quantity of our work. The excess free energy is

$$\Delta f^s = \frac{-1}{\beta A} ln(Z/Z_0), \tag{1.148}$$

where A is the area of the interface. We also deduce the fraction of occupied sites θ, [113]

$$\theta = -\frac{A\beta\lambda}{|\Lambda|} \frac{\partial \Delta f^s}{\partial \lambda}. \tag{1.149}$$

Eq.(1.147) shows that the SSM model maps the adsorption on a flat surface onto a two dimensional lattice problem of a very general kind, which is in general not amenable to analytic treatment. It can be simplified by introducing the Kirkwood superposition approximation

$$g_n^0(\mathbf{R}_1, \mathbf{R}_2, ..., \mathbf{R}_n) = \prod_{i<j\leq n} g_2^0(\mathbf{R}_i, \mathbf{R}_j), \tag{1.150}$$

where

$$g_2^0(\mathbf{R}_i, \mathbf{R}_j) = g_2^0(|\mathbf{R}_i - \mathbf{R}_j|) \tag{1.151}$$

is the pair correlation function.

The partition function can then be written as

$$Z/Z^0 = \sum_{\{t_i\}} e^{\beta\mu \sum_i t_i - \beta w \sum_{n.n.} t_i t_j} \qquad t_i = 0, 1, \tag{1.152}$$

where $\beta = 1/kT$,

$$\beta w = -\ln[g_2], \tag{1.153}$$

$$g_2 = g_2^0(\mathbf{R}_i, \mathbf{R}_j) \qquad [i, j = nearest \ neighbors], \tag{1.154}$$

$$\beta\mu = \ln[\lambda\rho_1^0(0)]. \tag{1.155}$$

The behaviour of the adsorbed film in this approximation depends on only two parameters, μ_i the adsorption affinity and w or more generally w_{ij}, the potential of mean force of two adsorbate moeties, i, j. The first quantity is obtained from the smooth wall approximation discussed in the previous section. The potential of mean force w determines the behaviour of the adsorbed film: If it is attractive, then first order phase transitions may occur. If it is repulsive then second order phase transitions can take place, such as order-disorder rearrangements. If the interaction is strongly repulsive not allowing first nearest neighbors on the triangular lattice then this becomes equivalent to the hard hexagon problem solved exactly by Baxter [59]

Further restriction of the interactions to nearest neighbors makes this problem equivalent to that of the Ising model with nearest neighbor interactions. Then the partition function can be mapped onto an Ising model with spin variables $s_i = \pm 1$ by means of the transformation

$$s_i = 2t_i - 1. \tag{1.156}$$

which has been solved exactly for several lattices[59].

Then there is a phase transition when $w < 0$,

$$\mu = wq/2 \tag{1.157}$$

or

$$\lambda\rho_1^0(0) = [g_2]^{-q/2}, \tag{1.158}$$

where q is the number of nearest neighbors of the lattice, 4 for the square lattice and 6 for the triangular lattice. The critical line for the first order phase transition for the triangular lattice [114] yields the expression

$$\theta = (1/2) \left(1 \pm \left(1 - \frac{16g_2}{(g_2 - 1)^3 (g_2 + 3)}\right)^{\frac{1}{8}}\right). \qquad (1.159)$$

Setting $\theta = (1/2)$, this equation yields the condition

$$g_2 \mid_{crit} = 3, \qquad (1.160)$$

and the value for the critical sticky parameter λ is

$$\lambda \rho_1^0(0) \mid_{crit} = 1/27. \qquad (1.161)$$

The contact pair correlation function in the bulk for ions of equal sign is practically zero, because of the Coulomb repulsion which prevents ions of equal sign to approach each other. In the adsorbed layer the interactions of these ions must be attractive, if the formation of a layer occurs suddenly, and therefore the state of chemical bonding, or more precisely the electrovalence, must change during the adsorption process.

When we turn off the interactions in the surface, then $w = 0$ and Eq.(1.152) becomes

$$Z/Z^0 = \sum_{\{t_i\}} e^{\beta \mu \sum_i t_i}, \qquad (1.162)$$

or

$$Z/Z^0 = (1 + e^{\beta \mu})^{|\Lambda|}, \qquad (1.163)$$

and using Eq.(1.149), we get

$$\theta = \lambda \frac{\partial}{\partial \lambda} \ln(1 + e^{\beta \mu}). \qquad (1.164)$$

Furthermore since

$$e^{\beta \mu} = \lambda \rho_1^0(0), \qquad (1.165)$$

we get the Langmuir isotherm

$$\theta = \frac{\lambda \rho_1^0(0)}{1 + \lambda \rho_1^0(0)}. \qquad (1.166)$$

The underpotential deposition of copper on gold

Phase transitions occuring during electrode processes have been studied using a model in which the electrode is a planar wall with sticky adsorption sites. This model is used to explain [115] of underpotential deposited films on perfect single crystal surfaces [116, 117] contain sharp spikes . In earlier work, we discussed the possibility of explaining these spikes as the result of first order phase transitions occuring in the surface [56, 57]. There are a number of conditions that have to be met to obtain experimentally sharp spikes in a voltammogram. These include chemical equilibrium, the degree of perfection of the substrate (a single crystal in most cases), and the scanning rate of the voltammogram. Ideally perfect single crystal surfaces with large domains, fast kinetics and diffusion should produce narrow spikes. Slow voltage scanning rates would be best to observe these spikes. The area under the spike is proportional to the change transferred, not to the coverage of the surface, because the charge per adatom on the surface is not necessarily an entire number, equal to the stoichiometric electrovalence [118]. However, sharp spikes are not the only interesting features of the voltammograms. New advances in ex and in situ surface analysis make it possible to determine and discriminate the origin of broadening effects. The case of the UPD, of Cu on Au(111) in the presence of H_2SO_4 has been extensively studied in recent times both experimentally [120, 20, 121, 1, 6, 11, 122] and also theoretically [123]. The picture that emerges shows that the broad foot observed in the voltammogram is not due neither to kinetic effects, nor domain size effects, nor to surface reconstruction, and is completely reversible. Therefore it is due to genuine statistical effect, that has to be accounted for. The broad foot of the first spike in the Cu-Au voltammogram can be explained by a second order surface phase transition, similar to the so called hard hexagon phase transition [59].

We assume that in the initial stages of the process there is a strong coadsorption of copper with the bisulfate. At positive potentials ($V > .4$ volts with respect of standard ($Ag/AgCl$) electrode, the bisulfate is strongly adsorbed onto the clean Au(111) surface. We assume, in accordance with chemical common knowledge, that it retains its charge, and therefore , the bisulfate-bisulfate interaction is both long ranged and repulsive. If we assume, as we have done in our previous work [58, 123], that the HSO_4^- sits in a tripod position, that is with its three oxygen atoms directly atop of the Au atoms of the surface, then the adsorption of one HSO_4^- necessarily excludes nearest neighbor occupation.

This makes the short ranged part of the surface interaction mathematically isomorphic to the hard hexagon problem, solved some years ago by Baxter[59]. According to this work, there will be a second order, order-disorder phase transition when

$$\theta_S \geq \theta_c = 0.2764. \tag{1.167}$$

where θ_S is the fraction of the Au(111) lattice adsorption sites that is occupied by the bisulfate.

Condensed phases in the ad layers are observed in electrochemistry. In particular the under potential deposition of some metals on electrodes occurs at certain very well defined values of the potential bias [121]. For example, the deposition of Cu on the Au(111) face forms two phases according to the deposition potential. These phases have been observed ex-situ [121] and in situ [20, 6, 11]. At a lower potential a dilute ordered phase is formed. At a higher potential a dense commensurate phase is formed. It is clear from the above considerations that in the dense ad layer case the ions must be discharged, because then they would form a metallic bond, which makes w negative, and therefore ferromagnetic. This is supported by the features of the EXAFS spectra. In the high density phase the near edge structure corresponds to that of metallic copper, which has a characteristic double peak.

In the electrosorption of ions, the charge of the ions can be neutralized by the electons in the metal electrode substrate. If this happens, then the normally repulsive effective interaction between equally charged ions can become attractive because of the formation of a metallic bond. The charge is known as the electrosorption valency, and has been studied extensively by Schultze and coworkers [118] and will be discussed in the next section. From the structure of Eq.(1.158) and Eq.(1.159) it is clear that no first order phase transition will occur if the adsorbed ions keep their charge and their repulsive interaction as the potential changes. The conclusion is that the addions, in this case Cu, attract each other in the adsorbed layer, and therefore are chemically different in the adlayer than in the bulk solution.

The contact density $\rho_i(0)$ is a function of the electric potential: an estimate of the contact density can be obtained using the expression used

$$\rho_i^0(0, \Psi) = e^{-\tilde{z}_i \Psi} \rho_i^0(0, 0), \tag{1.168}$$

where $\tilde{z}_i$ is the electrovalence and $\rho_i^0(0, 0)$ is the contact density of ion i. Ψ is the adimensional potential bias with reference to the potential of zero charge, and is given by

$$\Psi = \beta e[\psi(0) - \psi_{pzc}], \tag{1.169}$$

where e is the elementary charge, $\beta = 1/kT$, $\psi(0)$ is the potential at the electrode surface, and ψ_{pzc} is the potential of zero charge. Although Eq.(1.168) is certainly consistent with the classic Gouy-Chapman theory, and experimental evidence [124] shows that Eq.(1.168) holds even in systems where the assumptions of the Gouy-Chapman theory do not apply , and where we know that the discrete nature of the solvent will produce oscillatory charge profiles [29]. Yet contact theorems of the type

$$kT \sum_i \rho_i^0(0, \Psi) = \frac{\epsilon}{8\pi} E_0{}^2 + P_B, \tag{1.170}$$

where ϵ is the dielectric contant of the medium, E_0 is the bare electric field at the surface, and P_B is the bulk pressure, are valid for irregular surfaces with arbitrary interactions [119] in the mean field sense.

We consider the adsorption of a single ion only, always the counterion $i = 1$, which has a fugacity

$$z = \rho_i^0(0, \Psi)\lambda = e^{-\tilde{z}_1 \Psi} \rho_i^0(0, 0)\lambda. \tag{1.171}$$

The resummed series is just Langmuir's adsorption isotherm, which is indeed the correct physical limit of θ given by Eq.(1.166) when the lateral interactions are turned off. This immediately suggests the form of the Padé approximant for general z and $g_2 = 1/u$. For low fugacities

$$\theta_l = \frac{z + z(g_2 - 1)P(g_2, z)}{1 + z + z(g_2 - 1)P(g_2, z)}, \tag{1.172}$$

and for high fugacities

$$\theta_h = \frac{1}{1 + 1/[zg_2^6] + (1/[zg_2^6])(g_2 - 1)P(g_2, 1/[zg_2^6])}. \tag{1.173}$$

Using the dependence of z on the applied potential bias Ψ as given previously, it is possible to get the adsorption isotherms as a function of Ψ for fixed values of $\tilde{z}_1$ and $\rho_1^0(0, 0)\lambda$. For the case $\rho_1^0(0, 0)\lambda = 0.1$ and $\tilde{z}_1 = 1$ (monovalent counterion). A phase transition occurs when $y = 1/(zg_2^3) = 1$.

A phase transition will appear in the voltammogram as a sharp peak. The charge potential curve of the voltammogram can be obtained by differentiation:

$$I(\psi) = \frac{\partial \theta}{\partial \psi(0)} \frac{d\psi(0)}{dt}. \tag{1.174}$$

If the scanning rate is constant and we neglect diffusion and double layer effects [115], Eq.(1.169) and Eq.(1.174) yield

$$I(\psi) = -\tilde{z}_1 z \frac{\partial \theta}{\partial z} \frac{d\Psi}{dt}. \tag{1.175}$$

A final observation is that the mean field result can also be cast in the form of a modified Langmuir adsorption isotherm:

$$\theta = \frac{z g_2^{q\theta}}{1 + z g_2^{q\theta}},$$ (1.176)

Consider a model in which there are three species, which for brevity we will call E (empty sites), C (copper) and S (bisulfate). In our lattice model of the surface there is no interaction between E and the other adsorbates, S strongly repells S, and S-C and C-C are strongly attractive. Models in which three components are adsorbed where recently discussed in the literature [125, 126, 127, 113], but only nearest neighbor configurations were taken into consideration. In our case at least second neighbors need to be included.

Theory

The underpotential deposition (UPD) [116, 117] of metals frequently involves phase transitions that are observed in the voltammograms as sharp spikes [128, 30]. In earlier work [56, 57] we introduced a sticky adsorption site model to study phase transitions that occur when the adsorbate is commensurate with the substrate. This situation is known to occur, for example in the UPD of copper on gold (111) in the presence of bisulfate ions. As has been shown in earlier work [58, 129, 130] the particular structure of this voltammogram is due to a sequence of first and second order transitions taking place on the surface. To obtain sharp spikes in a voltammogram the substrate surface must be a perfect single crystal with large domain size. Slow scanning rates are also required. This immediately suggests that thermodynamic equilibrium and reversibility are necessary for the occurence of these spikes, which are associated with first order phase transitions in the adsorbate layer [56, 57].

We should mention in this context the early work of Bewick and Thomas [131] who studied the case of UPD of Pb on Ag (111), and the work of Buess-Hermann[132] on the adsorption of alcohols on mercury. In both cases the mean field Frumkin's isotherm [133] was used to interpret the observations. We observe here that in both cases the adsorbed layer is incommensurate with the substrate, and in our theory the adsorbate must be commensurate. For the commensurate case Frumkin's isotherm corresponds to the mean field theory, which is known to be very inaccurate. The coexistence curve is of the form $y = x^{1/2}$ for Frumkin, while the exact lattice result is $y = x^{1/8}$ [134]. This has important consequences for the shape of the curves and for the sensitivity of the voltammogram to the interaction parameter of the adsorbate. In the case of the exact lattice gas isotherm the occupancy θ_{Cu} changes suddenly from an almost Langmuir type behaviour to a phase transition at the critical value of the interaction parameter g_2. In the mean field theory the change is much more gradual.

For the incommensurate case, or for the liquid mercury electrode, Frumkin's isotherm describes a gas-liquid transition, and not the solid-liquid transition [135]. This opens interesting possibilities about different types of transitions that are possible, and that we hope to discuss in the future. In the voltammogram the area under the spike is proportional to the charge transferred, not to the coverage of the surface, because the charge per adatom on the surface is not necessarily an entire number equal to the stoichiometric electrovalence [118].

The model

The case of the UPD of Cu on Au(111) in the presence of H_2SO_4 has been extensively studied in recent times both experimentally [136, 20, 137, 138, 120, 121, 1, 139, 140, 122] and theoretically [123]. We should also mention that two recent studies of the kinetics of this system [141, 142] clearly indicate that in the shape of the voltammogram the first peak can be derived from equilibrium considerations, as was done in our earlier work, while in the second the kinetics is much slower and therefore observable from the analysis of the voltammogram.

In recent papers [58, 129] a model for the underpotential deposition of Cu on Au(111) in the presence of bisulfate ions was proposed. We summarize here the main results of our previous work.

In our model for the underpotential deposition of Cu on Au(111) in the presence of sulfuric acid, we assume that a well defined sequence of events takes place:

- Bisulfate ions are adsorbed at very positive potential, forming a $\sqrt{3} \times \sqrt{3}$ lattice on the gold surface. These bisulfate ions are desorbed as the potential is decreased, undergoing a hard hexagon like second order phase transition.

- Copper ions are then adsorbed on the free adsorption sites. The adsorption of copper produces a reabsorption of the bisulfate ions, which eventually will undergo the hard hexagon transition to rebuild the $\sqrt{3} \times \sqrt{3}$ lattice on the gold surface. This forms a honeycomb lattice for the adsorption of the remainder of the copper of the first peak.

- In the second peak the adsorbed bisulfate ions are displaced by copper ions from the $\sqrt{3} \times \sqrt{3}$ positions. However, they could still remain bound to the copper, which now forms a full monolayer on top of the Au(111) surface.

We assume that the bisulfate binds to the gold (111) surface in such a manner that the sulfur is directly on top of the adsorption site for the copper, three of the bisulfate oxygens being above and directly associated with the three gold atoms of the surface, which form a triangle about the adsorption site. Packing considerations indicate that two bisulfate groups cannot be adsorbed onto neighboring adsorption sites. The bisulfate ions will thus form a $\sqrt{3} \times \sqrt{3}$ film by occupying one of the three triangular sublattices Λ_T of the full triangular lattice of adsorption sites, with a maximum coverage of 1/3.

Bisulfate adsorption

In our model it was assumed that the bisulfate ion formed a $\sqrt{3} \times \sqrt{3}$ template. This template leaves a honeycomb lattice of free sites for the adsorption of copper. The clear implication is that the first peak has 2/3 of a monolayer of Cu. The second peak corresponds to the replacement of the bisulfate by copper in the adlayer. We showed also that the broad foot of the first peak is due to a second order hard hexagon like transition, which is seen experimentally by Itaya[136] and Kolb [143]. We believe that the interpretation that the first peak corresponds to only 1/3 of a monolayer, based on the STM and LEED observations, is consistent with our model if it is the bisulfate ion that is seen.

In our model [58, 129] the broad foot of the first spike in the Cu-Au voltammogram is due a second order surface phase transition, similar to the hard hexagon phase transition [59]. This transition was actually observed by Itaya [136] and by Kolb [143].

In our theory in the initial stages of the process there is a strong coadsorption of copper with the bisulfate. At positive potentials ($V > 0.45$ volts with respect to the standard (Ag/AgCl) electrode), the bisulfate is strongly adsorbed onto the clean Au(111) surface. This means that there is an ordered $\sqrt{3} \times \sqrt{3}$ structure for these very positive potentials observable by STM or AFM. We assume that it retains its charge, and therefore, the bisulfate-bisulfate interaction is both long ranged and repulsive. The HSO_4^- sits in a tripod position, that is with its three oxygen atoms almost directly atop the Au atoms of the substrate, so that the adsorption of one HSO_4^- necessarily excludes nearest neighbor occupation. This makes the short ranged part of the surface interaction mathematically isomorphic to the hard hexagon problem, solved mathematically by Baxter[59]. There will be a second order, order-disorder phase transition when

$$\theta_S \geq \theta_c = 0.2764, \tag{1.177}$$

where θ_S is the fraction of the Au(111) lattice adsorption sites that are occupied by the bisulfate.

Consider a model in which there are three species, which for brevity we will call E (empty sites), Cu(copper) and S (bisulfate). In our lattice model of the surface there is no interaction between E and the other adsorbates, S strongly repels S, and S-Cu and Cu-Cu are strongly attractive. Models in which three components are adsorbed were recently discussed in the literature [125, 126, 127, 113].

We define [56, 57] the inner layer equivalent fugacity z_S for the adsorption of the bisulfate

$$z_S = \lambda_S(\psi)\rho_S^0(0, \psi) \tag{1.178}$$

where the sticking coefficient can be interpreted as $\lambda_S(\psi) = \exp[\beta\mu_S]$, with μ_S as the free energy change that occurs when a bisulfate ion binds to the gold surface, removing in the process the adsorbed water. $\rho_S^0(0, \psi)$ is the inner layer local density of bisulfate for a local potential ψ.

We assume the simple exponential form [115] for the fugacity

$$z_S = \lambda_S^0 \rho_S^0(0,0)e^{-\zeta_S \beta e(\psi - \psi_S^{Re})} \tag{1.179}$$

where $\beta = 1/kT$ is the Boltzmann thermal factor, the electrosorption valency of the bisulfate is $\zeta_S = -1$, and ψ_S^{Re} is the electrosorption reference potential, that depends on the nature of the substrate.

The adsorption isotherm for the hard hexagon model has been derived by Joyce [144]. His results can be fitted to Padé approximants [129]. For the high density, ordered phase

$$\theta_S^{high} = \frac{0.2764 + 0.155(z_S - 11.09) + 0.01(z_S - 11.09)^2}{1 + 0.5(z_S - 11.09) + 0.03(z_S - 11.09)^2} \qquad z_S > 11.09. \tag{1.180}$$

For the low density, disordered phase,

$$\theta_S^{low} = \frac{0.709z_S + 0.0079z_S^2 + 0.0011z_S^3}{1 + 3.3z_S} \qquad 0 < z_S < 11.09. \tag{1.181}$$

Consider now the adsorption of bisulfate onto Au(111) in the absence of copper. Using Eq.(1.179) with the effective charge of the adsorbed bisulfate $\zeta_S = -1$,

$$z_S = \lambda_S^0 \rho_S^0(0,0)e^{\beta e(\psi - \psi_S^{Au})} = z_S^0 e^{\beta e(\psi - \psi_S^{Au})} \tag{1.182}$$

where we take

$$\psi_S^{Au} = 0.388V \tag{1.183}$$

and the bisulfate fugacity z_S is ($z_S^0 = 1$, $\qquad T = 298.16K$)

$$z_S = e^{38.922[\psi - \psi_S^{Au}]}, \tag{1.184}$$

The bisulfate is desorbed as the potential turns less positive. With these parameters the S lattice becomes disordered at about 0.45 V, which is when the copper starts to deposit. As long as the S template is disordered, the Cu is adsorbed randomly (probably next to the bisulfate).

The copper will start depositing at around $\psi = 0.4V$, but since the bisulfate is in its disordered phase, there should be no cooperative effects due to the copper-copper attraction on the surface, and therefore, to a first approximation we may assume that the electrodeposition of copper follows a Langmuir-like (or Frumkin-Langmuir [115]) adsorption isotherm

$$\theta_{Cu} = \frac{\hat{z}_{Cu}}{1 + \hat{z}_{Cu}} \qquad \theta_S < \theta_S^{crit} = 0.2764 \tag{1.185}$$

where ψ_{Cu}^{Au} is the electrodeposition reference potential of Cu on gold (111), and ζ_{Cu} is the effective electrovalence (certainly related to the electrosorption valency) of Cu (see the discussion below, Eq.(1.203)).

The electrosorption of bisulfate on polycrystalline copper [145, 146] shows that it binds much more strongly than to gold. We may assume that the electrosorption potential of the bisulfate varies linearly with the amount of copper in the surface. Thus, guided by the work of Trasatti on the influence of the anion on the potential of zero charge [147], we take

$$\psi_S^{Cu} < \psi_S^{Au}. \tag{1.186}$$

In our model, then

$$z_S = z_S^0 e^{-\zeta_S \beta e(\psi - \psi_S)}, \tag{1.187}$$

with

$$\psi_S = \psi_S^{Cu}\theta_{Cu} + (1 - \theta_{Cu})\psi_S^{Au}, \tag{1.188}$$

In the presence of copper the bisulfate is then readsorbed. Eventually, the fraction of occupied sites will again surpass the critical value of 0.2764 (which corresponds to 83% of the occupancy of the template $\sqrt{3} \times \sqrt{3}$ lattice), and then there will be a first order transition for the copper on the honeycomb lattice of the Cu sites.[148] If we assume that the occupancy of copper $\theta_{Cu} = 1/3$ at the transition, we get that $\psi_S^{Cu} = -.302V$, which is in qualitative agreement with Wieckowski [145, 146].

The fugacity of copper

We recall that in the using the sticky site model [56, 57], there are two basic parameters that determine the phase behaviour of the adsorbed layer:

- The local fugacity $\hat{z} = \hat{z}_{Cu}$ of the two dimensional adsorbate gas is the product of the sticking probability λ_{Cu} and the contact density $\rho_{Cu}^0(0, \psi)$ of the ion i,

$$\hat{z}_{Cu} = \lambda_{Cu}\rho_{Cu}^0(0, \psi) \tag{1.189}$$

- The lateral interaction parameter g_2

$$g_2 = g_2^0(\mathbf{R}_i, \mathbf{R}_j) \qquad [i, j = nearest \; lattice \; neighbors], \tag{1.190}$$

where $\mathbf{R}_i, \mathbf{R}_j$ are the positions of the neighboring lattice sites i, j, and $g_2^0(r)$ is the pair correlation function in the electrode plane, but in the absence of the adsorption sites.

Both of these parameters depend on the applied potential ψ. Consider first the fugacity: The sticking probability is a function the bonding free energy ε_{Cu} of the adsorbate to the binding site. This quantity is independent of the ionic strength of the electrolyte, for a given potential. However, it will depend on the potential bias [115] at the surface. We write

$$\lambda_{Cu} = \lambda_{Cu}^0 e^{\beta \varepsilon_{Cu}}, \tag{1.191}$$

where the binding energy of the Cu ion depends on the potential bias of the metal substrate, or electrode,

$$\varepsilon_{Cu} = \zeta_{Cu} e[\psi - \psi_{Cu}^{Re}], \tag{1.192}$$

where e is the elementary charge, ψ is the potential at the electrode surface, and ψ_{Cu}^{Re} is the electrosorption reference potential. ζ_{Cu} is the effective charge of the adsorbed species at the surface, not necessarily equal to the electrovalence in solution.

The contact density $\rho_{Cu}^0(0, \psi)$ is a also function of the potential bias. As a first approximation we will assume that the contact density is given by

$$\rho_{Cu}^0(0, \psi) = \rho_{Cu}^0(0, 0)e^{-\beta e\nu_{Cu}\psi} \tag{1.193}$$

where $\rho_{Cu}^0(0, 0)$ is the contact density at zero potential ψ, which for the Gouy-Chapman theory is equal to the bulk density ρ_{Cu}, and $e\nu_{Cu}$ is the full charge of the Cu ion. Therefore

$$\hat{z}_{Cu} = \lambda_{Cu}^0\rho_{Cu}^0(0, 0)e^{\beta e[(\varsigma_{Cu}-\nu_{Cu})\psi-\varsigma_{Cu}\psi_{Cu}^{Re}]} \tag{1.194}$$

Similarly, we write for the lateral interaction parameter g_2

$$g_2 = e^{-\beta\omega_2}. \tag{1.195}$$

where ω_2 is the potential of mean force between moeties adsorbed in neighboring sites. Since we know that the charges of these moeties change with ψ[39, 40], ω_2 must also be a function of the potential bias ψ. (A detailed theory of a model which calculates this effect will be presented separately.)

Therefore we may expand about ψ_{Cu}^{Re}

$$\omega_2 = \omega_2^0 + e\alpha[\psi - \psi_{Cu}^{Re}] + ... \tag{1.196}$$

where

$$\alpha = \frac{1}{e}\left|\frac{\partial\omega_2}{\partial\psi}\right|_{\psi=\psi_{Cu}^{Re}} \tag{1.197}$$

As was discussed in earlier work [56, 57], the coexistence curve is given by the exact condition

$$\hat{u} = \hat{z}[g_2]^{q_L/2} = 1, \tag{1.198}$$

where q_L is the number of neighbors of the lattice, 6 for the triangular (center filled regular hexagons) lattice and 3 for the honeycomb (empty center hexagons). If the parameter $\hat{u}$ is less than unity then we are in the 1 phase region. If it is larger than 1, then we are in the two phase region. The coexistence curve is obtained setting $\hat{u} = 1$ in Eq.(1.198). For the triangular lattice [114]

$$\theta_T = (1/2)\left(1 \pm \left(1 - \frac{16g_2}{(g_2 - 1)^3 (g_2 + 3)}\right)^{\frac{1}{8}}\right), \tag{1.199}$$

with a critical value $g_2^T = 3$. For the honeycomb lattice [148]

$$\theta_H = (1/2)\left(1 \pm \left(1 - \frac{16g_2^{3/2}(1 + g_2^{3/2})}{\left(g_2^{1/2} - 1\right)^3 (g_2 - 1)^3}\right)^{\frac{1}{8}}\right), \tag{1.200}$$

with a critical value $g_2^H = (2 + \sqrt{3})^2 \simeq 13.93$.

As it was shown in earlier work, [57] the adsorption isotherm can be expressed as a generalized Langmuir adsorption isotherm of the form

$$\theta_{Cu} = \frac{\mathcal{A}(\hat{z}, \hat{u})}{1 + \mathcal{A}(\hat{z}, \hat{u})}, \tag{1.201}$$

where $\mathcal{A}(\hat{z}, \hat{u}) = \mathcal{A}$ is a polynomial in $\hat{z}$ and $\hat{u}$. For the case in which there is a phase transition, θ_{Cu} is practically a step function which was represented by an error function in our previous work. A very simple and useful alternative form is

$$\mathcal{A} = \hat{z} + (g_2 - 1)\hat{u}^n, \tag{1.202}$$

where n is some entire number. The width of the transition spike is inversely proportional to n (actually to $n/4$) , and will play the same role as the width parameter of our previous work[149]. It is related to instrumental width, substrate domain size and defects on the surface.

Combining Eq.(1.194)- Eq.(1.196) and Eq.(1.198), we get that at the transition point, ψ_T

$$\hat{u}_{Cu} = \lambda_{Cu}^0 \rho_{Cu}^0(0,0) e^{\beta e[(\zeta_{Cu} - \nu_{Cu} - q_L \alpha/2)\psi_T - (\zeta_{Cu} - q_L \alpha/2)\psi_{Cu}^{Re} - q_L \omega_2^0/2e]} = 1. \tag{1.203}$$

The relation of the parameters in this formula to the electron transfer coefficients will be discussed in future work.

The current in the voltammogram

Consider the electrode interface, in which a current flows. We have specifically in mind the case of the UPD of Cu onto Au(111). In this case there are two ions that participate in the transport of charge:

- The bisulfate, that carries a negative charge which it keeps when adsorbed onto the surface. Therefore, it will contribute only to the capacitive current density j_C. The other contribution to j_C is from the diffuse layer.

- The copper, which is adsorbed and discharged to some extent. It's electrovalence will change from ν_{Cu} in the bulk phase to ζ_{Cu} at the electrode interface. Furthermore, ζ_{Cu} should be a function of the potential ψ.

Therefore, the total current density is (for a recent discussion see, for example the work of De Levie [150] and Lantelme [151])

$$j_T = j_C + j_F \tag{1.204}$$

where j_T is the total current density, j_C is the capacitive contribution and j_F is the Faradaic contribution. If we neglect double layer effects, then the capacitive current is due to the bisulfate, and the faradaic current almost exclusively to the discharge of the copper ions. Then

$$j_C = (1/A) \left[C_i + \psi \frac{dC_i}{d\psi} \right] \frac{d\psi}{dt} \tag{1.205}$$

gives the current due to the discharge of the capacitor as well as its change in integral capacitance C_i. A is the area of the electrode.

The current associated with the cation (in this case the copper) can be written as

$$j_F = (M/A)e \left[-\theta_{Cu} \frac{d\zeta\_Cu}{d\psi} + (\nu_{Cu} - \zeta_{Cu}) \frac{d\theta_{Cu}}{d\psi} \right] \frac{d\psi}{dt} \tag{1.206}$$

where M is the number of adsorption sites per area A, e is the elementary charge, ζ_{Cu} is the actual charge of the adsorbed copper and ψ the potential.

In accordance with the quantum theory [39, 40], the charge of the adsorbate will change with the applied potential because the electron density at the surface is changing. Intuitively we take the exponential form

$$\zeta_1(t) = e^{\omega_\zeta (\psi - \psi_1^{Re})}, \tag{1.207}$$

where ω_ζ and ψ_1^{Re} are adjustable parameters.

Geometry of the model

We discuss now some geometrical properties of our model. We recall that we assumed that the bisulfate ion formed a $\sqrt{3} \times \sqrt{3}$ template. This template leaves a honeycomb lattice of free sites for the adsorption of copper. The clear implication is that the first peak has 2/3 of a monolayer of Cu. In this section we want to show that the bisulfate actually protrudes from the surface, and therefore the 1/3 of a monolayer seen by the STM and LEED observations is consistent with our model.

The vertical distance between the substrate Au(111) plane and the adsorbed copper can be estimated to be 2.22 Å , by taking the mean distance in metallic gold and in metallic copper. From early EXAFS experiments [20, 137, 138] and recent theory [123] we know that the gold (or copper) distance to the oxygens in the bisulfate is between 1.95Å and 2.1 Å , so that from the dimensions of the bisulfate ions, it should be between 1.84Å and, if we count the hydrogen of the bisulfate, 2.13Å higher than the plane where the copper lies. Assuming the same geometry on a clean Au(111) surface the bisulfate layer should stand about 4.06-4.35 Å above the Au surface. However, the STM cannot measure absolute heights, and therefore, both cases appear as a $\sqrt{3} \times \sqrt{3}$ overlayer.

An observation to be made is that the $Cu - O$ distances appear at 1.81 Å if we assume an undistorted bisulfate ion and copper honeycomb lattice. This is a bit lower than the expected value of 1.95Å . Therefore we would expect that the adsorbed ion be deformed so that the oxygens lie about .2 Å below the line of the copper atoms, which would relieve the stress, but at the same time create a activation energy barrier for the desorption of the copper. This again is consistent with the experimental observations.

The proposed coordinates are shown in Table 1.1.

Concentration dependence and electrovalence

If all the parameters stay the same as the concentration of copper is changed, a shift in the concentration will necessarily imply a shift in the transition potential

$$\Delta \psi_T^{Re} = \frac{kT}{e(\nu_{Cu} + q_L \alpha/2 - \zeta_{Cu})} \Delta \ln[\rho_{Cu}(0,0)]. \tag{1.208}$$

For a tenfold increase in concentration the potential shift is

$$\Delta \psi_T = \frac{59.2 mV}{\nu_{eff}}, \tag{1.209}$$

where ν_{eff} is the effective charge obtained directly from the equation Eq.(1.203)

$$\nu_{eff} = \nu_{Cu} + q_L \alpha/2 - \zeta_{Cu}. \tag{1.210}$$

	undistorted-stressed			distorted-relaxed		
	x	y	z	x	y	z
Au1	0	-1.66	0			
Au2,3	±1.44	0.88	0			
Cu1,2	±2.88	0	2.22			
Cu3,4,5,6	±1.44	±2.49	2.22			
O1	0	-1.39	2.09	0	-1.32	1.9
O2,3	±1.21	0.70	2.09	±1.15	0.66	1.9
O4	0	0	4.065			
S	0	0	2.58			
H	0	0.90	4.35			

Table 1.1: Structure of the adlayer with 1/3 coverage of bisulfate and 2/3 of Cu. The stressed structure was made using an undistorted bisulfate ion. In the relaxed structure the bisulfate ion was squeezed so that the $O - O$ distance is shortened from 2.42 Å to 2.30 Å , while the copper- oxygen goes from 1.81 to 1.87 Å.

We are now in position to discuss the recent experiments of Omar et al. [142] and Hölzle et al [141]. Using the results of the local kinetic theory [153] we can estimate the positions of the peaks in the experiment of Omar et al. We notice that the relative shift of the bulk deposition edge corresponds exactly to the electrovalence of copper, 2. We should remark that this a bit surprising, since one would expect that it should scale to the activity of the copper, not its concentration. However this also means that the contact density $\rho_{Cu}(0, \psi)$ in our model scales almost exactly to the Gouy- Chapman estimate Eq.(1.193), which is nontrivial and reassuring. We recall that our interface is not planar, our solution is molecular, and the position at which $\rho_{Cu}(0, \psi)$ is taken is some average over the positions in the metal electrolyte interface. We know that the contact theorems are satisfied by these averages, and this is probably the reason why the Gouy-Chapman estimate is so accurate.

If we use the shifts of the edge to compute the effective charge Eq.(1.210) for the peaks in the voltammogram we find that the effective charge for the second peak is 20 % higher than 2, and that for the first peak we get about 1, which is very difficult to explain in terms of individual fugacities or the Nernst equation [142].

The inclusion of the variation of the pair correlation functions between adsorbed particles with ψ removes these difficulties. The results of table 1.2 show that the remarkable accuracy of the Gouy-Chapman estimate holds also for other transition points. However the actual values of the charges cannot be reconciled with the other experimental evidence.

For the second peak a charge bigger than two cannot be explained in any other way but by including the variation of g_2 with potential ψ. The value $\alpha = 0.14$ seems reasonable.

The apparent charge for the first peak $\nu_{eff} = 1$ is definitely inconsistent with the experimental evidence and in particular with the integrated charge, where the ratio of the charges of both peaks is very closely 1, so that each peak corresponds to 50% of the monolayer charge. If we assume only 1/3 coverage by copper as has been indicated in the literature [136, 143], the charge would be only 16% of the monolayer, clearly impossible. The explanation that the remainder of the charge is due to desorption of the sulfate is inconsistent with the EXAFS experiment [20], with the radiotracer experiment[145], and with the microbalance experiment, all of which indicate that there is strong bisulfate adsorption even after the full monolayer is deposited.

The situation improves if we take the 2/3 monolayer coverage predicted theoretically [58, 129]. We get 50 % of the monolayer for each peak's charge when $\nu_{Cu} - \zeta_{Cu} = 1.0$ and $\zeta_{Cu} = 1.0$. The value $\zeta_{Cu} = 0.5$, which implies $\nu_{eff} - \zeta_{Cu} = 1.5$ satisfies the 60% for the first peak charge requirement of Kolb's votammogramm, and produces a neutral film near the expected value of the point of zero charge for this system. We believe that this last one is a more plausible value of the charge, since charge neutrality explains the lifting of the reconstruction of the gold surface by the adsorbed copper layer.

| | Exp.(mV) | | | Shifts(mV) | | e | | |
	#1	#2	#3	#2→#1	#3→#2	Eff.Chg.	Real.Chg.	α
Edge	299	270	241	29	29	2.04	0	
Foot	557	500	458	57	42	-		
1st. peak	554	496	444	58	52	1.02/1.14	0.5	-0.33/-0.25
2st. peak	352	327	303	25	24	2.37/2.47	0	0.14

Table 1.2: Shifts of the voltammogram's peaks with concentration from Omar et al [142]. **Experiment # 1** : $9.10^{-2}H_2SO_4$, $5.10^{-2}CuSO_4$, **Experiment # 2** : $9.10^{-2}H_2SO_4$, $5.10^{-3}CuSO_4$, **Experiment # 3** : $9.10^{-2}H_2SO_4$, $5.10^{-4}CuSO_4$. The effective charge is evaluated using Eq.(1.210). α is evaluated using $\zeta_{Cu} = 0.5$ for the first peak, and $\zeta_{Cu} = 0$ for the second peak. The foot corresponds to the hard hexagon transition, which we cannot estimate accurately from the graphs.

1.4   Quantum mechanical treatment

The quantum mechanical treatment of the metallic surface has been developed to a large extent for metal-vacuum interfaces. Similar techniques should be applicable to the metal electrolyte interface. The main difference is that the fluid contributes not only to the electric fields in the metal surface, but also to specific chemical bonding interactions.

It is important in the quantum mechanical treatment of the electrode interface, therefore, to have a good knowledge of the structure of the electrolyte near the electrode, and also, of the electric fields at the interface.

1.4.1   Quantum mechanical treatment of the interface

The quantum theory of the metal interface with vacuum has been well developed. Most of the succesful treatments are based on functional density theory [152, 154, 155]. General reviews are available [156, 157]. The metals of interest in electrochemistry are the noble metals and mercury. These are sp and d band metals. The valence electrons of an sp-metal are delocalized and can be considered as a Fermi fluid interacting with the lattice of metal ions. In the jellium model, the interaction of the electrons with the ions is simplified by smearing out the positive ionic charge into a background of constant density, which terminates at the metal surface. Since in this average the information about the structure of the ionic lattice is lost, the jellium model is applicable to liquid metals or to polycrystralline surfaces, but it is still a useful model. The electron cloud spills over outside the metal surface, and the extent of this spill depends on the potential drop between the metal surface and the bulk electrolyte. The results of Lang and Kohn [155] are particularly interesting.

The electronic density $n(x)$ satisfies the Budd-Vannimenus sum rule [158], which is the quantum mechanics version of the dynamic balance equation

$$n(-\infty)[\phi(0) - \phi(-\infty)] = -p(-\infty) - 2\pi q_s^2 \qquad (1.211)$$

where $\phi(z)$ is the electrostatic potential, $p(z)$ is the pressure. The bulk jellium pressure is obtained from Wigner's equation

$$p(-\infty) = (k_F^2/2) \left[0.4 - 0.0829 r_s - \frac{0.0796 r_s^3}{(r_s + 7.8)^2} \right] \qquad (1.212)$$

where

$$k_F = (3\pi^2 n_+)^{1/3}$$

is the Fermi momentum and

$$r_s = [3/(4\pi n_+)]^{1/3}$$

is the Wigner-Seitz radius. The electron density can be parametrized as

$$n(z) = n_+[1 - Ae^{\alpha z}\cos(\gamma z + \delta)]\theta(-z) + n_+ Be^{-\beta z}\theta(z) \tag{1.213}$$

which gives a fair representation of the density profile as computed by more accurate methods [155]. Of the six parameters of this equation, four can be obtained from the Budd Vanneminus sum rule, charge balance, and the continuity of the electronic density $n(z)$ and its derivative at $z = 0$.

The density profile changes as the potential is changed. If we plot the charge outside of the metal B as a function of the potential difference $[\phi(0) - \phi(-\infty)]$. We find that the electron density spill increases with the potential bias.

1.4.2 Electrosorption on metal electrodes

It was first recognized by Lorenz and Salie [159] that when an ion is adsorbed on an electrode the charge that is transferred need not be a whole multiple of the unit charge. This partial charge transfer can be represented by the stoichiometric equation

$$S^z + \lambda e^- = S_{ad}^{z-\lambda} \tag{1.214}$$

There is no direct way of measuring the partial charge transfer λ, the most direct experimental evidence comes from NEXAFS experiments. It is related to the electrosorption valency [160].

We consider a nonadsorbing suporting electrolyte of high concentration, so that the charges in the interfacial region are well screened. Then, we define the electrosorption valency as

$$\gamma_i = \frac{1}{F}\left[\frac{\partial \tilde{\mu}_i}{\partial \Delta\varphi}\right]_{\Gamma_{ad}} \tag{1.215}$$

where $\tilde{\mu}_i$ is the electrochemical potential of i, $\Delta\varphi$ is the potential drop, Γ_{ad} is the surface concentration of the adsorbate, and F is Faraday's constant. The electrosorption valency is the surface analog of the Nernst valency for the reaction

$$\sum_i \nu_i S_i + ne^- = 0 \tag{1.216}$$

for which

$$\nu_i/n = \frac{1}{F}\left[\frac{\partial \tilde{\mu}_i}{\partial \Delta\varphi}\right]_{\tilde{\mu}_{j\neq i}} \tag{1.217}$$

and we identify ν_i/n to the electosorption valency γ_i

Consider the surface Gibbs Duhem relation

$$\mu d\Gamma + \Gamma d\Delta\varphi + \ldots = 0 \tag{1.218}$$

that in the presence of a supporting electrolyte

$$\gamma = \frac{1}{F}\left[\frac{\partial q}{\partial \Gamma_{ad}}\right]_{\Delta\varphi} \tag{1.219}$$

from where

$$j = \frac{dq}{dt} = -\gamma F\left[\frac{\partial \Gamma_{ad}}{\partial t}\right]_{\Delta\varphi} \tag{1.220}$$

so that clearly the electrosorption valency is related to the charge transfer. In order to obtain a precise relation we need to introduce a non thermodynamic quantity, the geometric factor g

$$g = \frac{\phi_{ad} - \phi_s}{\phi_m \phi_s} \tag{1.221}$$

where $\phi_m = \phi(-\infty)$ is the potential in the metal, $\phi_s = \phi(\infty)$ is the potential in the solution, and ϕ_{ad} is the potential at the adsorbed layer, which is the non thermodynamic quantity. At the potential of zero charge we get

$$\gamma(0) = zg + \lambda(1-g) + \mu_{ad} - \nu\mu_w \tag{1.222}$$

where μ_{ad} is the dipole moment of the adsorbate, μ_w that of water, z is the electrovalence, of which λ is transferred to the electrode according to Eq.(1.214). Thus, the charge transfer is related to the electrosorption valency by non thermodynamic quantities.

Electrostatic potential at a metal/electrolyte interface gives the effect of the ionic charge number, z, which is transported over a fraction g of the potential difference $\Psi_m - \Psi_a$

1.4.3 Theoretical discussion of the electrosorption valency

Consider the case where only one valence orbital is involved in the adsorption bond. We are interested in the energy of this level relative to the Fermi level of the metal, which will determine to which extent the energy levels are occupied.

Since we are interested in equilibrium phenomena, fluctuations, and in particular, those of the solvation layer can be neglected.

To determine the change in the free energy, ΔE_b , required to transfer an electron from of the metal to the ion in the bulk of the solution. We decompose the electron transfer process from the metal into separate steps:

$$
\begin{aligned}
e^-(metal) &\rightarrow e^-(outside - metal) & \Delta G &= \Phi \\
e^-(outside - metal) &\rightarrow e^-(outside - solution) & \Delta G &= -e_0[\Psi_s - \Psi_m] \\
S^z(solution) &\rightarrow S^z(outside - solution) & \Delta G &= -\Delta G^f_{sol}(S^z) \\
S^z + e^-(outside - solution) &\rightarrow S^{z-1}(outside - solution) & \Delta G &= I \\
S^{z-1}(outside - solution) &\rightarrow S^{z-1}(solution) & \Delta G &= -\Delta G^f_{sol}(S^{z-1})
\end{aligned}
$$

Here "outside" means a position which is close enough to the phase (metal or solution) that potential is equal to the outer potential but which is far enough so that image forces are negligible. Φ is the work function of the metal, Ψ_s and Ψ_m are are the outer potentials of the solution and the metal, respectively. $\Delta G^f_{sol}(S^z)$ is the freee energy of solvation of S^z, which is known experimentally. Finally I is the ionization energy of the solute, generally not measured directly and e_0 is the electron charge. The total change in free eenrgy is

$$\Delta G^b = \Phi - e_0[\Psi_s - \Psi_m] - \Delta G^f_{sol}(S^z) + \Delta G^f_{sol}(S^{z-1}) - I \tag{1.223}$$

As an ilustration, we consider Cs^+ and Cl^- in aqueous solution near a polycrystalline gold electrode. We assume that both the metal and the solution are at the same outer potential. Then we get for Cs^+ [161]

$$\Delta G^b = 4.80 + 2.94 - 3.89 = 3.85 eV$$

For Cl^-

$$\Delta G^b = 4.80 - 3.07 - 3.61 = -1.88 eV$$

These values merely confirm the well known fact that both ions are stable in the neighborhood of gold at the point of zero charge.

1.4.4   Model Hamiltonian

Electrosorption is similar to adsorption from the vacuum, and can be treated using the same techniques. The presence of the solution has to be included, which necessarily complicates the calculations. The Anderson-Newns model (for a review, see Refs [162, 163]), which is the standard theory for adsorption on the metal-vacuum interface has been extended to the etal solution interface [164, 165]

We follow the treatment of Kornyshev and Schmickler [165]. We consider a single atom adsorbed at a metal/electrolyte interface. Only one adatom orbital interacts with the metal, therefore, all collective interactions are neglected, and in particular, all surface phase transitions. We denote the adatom electronic energy by ϵ_a The total Hamiltonian of the system is composed of three terms,

$$H_t = H_a + H_m + H_s \tag{1.224}$$

The first term corresponds to the valence electrons of the adatom

$$H_a = \sum_\sigma [\epsilon_a n_{a\sigma} + U n_{a\sigma} n_{a-\sigma}] \tag{1.225}$$

Here U is the Coulomb repulsion energy of two valence electrons, σ is the spin occupation number operator, and the index "a" stands for adsorbate.
The interaction of the metal electrons and the adsorbate is

$$H_m = \sum_{k\sigma} [\epsilon_k n_{k\sigma} + V_k (c_{k\sigma}^+ c_{a\sigma} + h.c.)] \tag{1.226}$$

where k denotes a set of quantum numbers for the metal electrons, V_k is the matrix element for the electron exchange exchange between the metal and the adsorbate, and c^+ and c denote creation and anhilation operators for the states indicated by the subindex. h.c. denotes hermitian conjugate.

In considering the adatom-solvent interaction, we follow the theories of electron transfer reactions in solution (for a review, e.g., Ref.[166]) and distinguish between the librational and vibrational modes of the solvent, which are slow compared to the rate of electron exchange, and the polarizability of the solvent molecules. In the harmonic approximation, we can write for the slow solvent modes and their interaction with the adsorbate in the following expression

$$H_s = 1/2 \sum_\nu \hbar \omega_\nu (p_\nu^2 + q_\nu^2) + (z - \sum_\sigma n_{a\sigma}) \left(\sum_\nu \hbar \omega_\nu g_{a\nu} q_\nu \right) \tag{1.227}$$

Here, the ν labels are for the slow solvent modes of frequency ν , momentum p_ν and coordinate q_ν. $g_{a\nu}$ is the coupling constant for the interaction with the adsorbate, which couples electronic energy to the solvent coordinates q_ν . Instead of using previous models, one can also model the solvent as an ensemble of hard dipoles.

To calculate the equilibrium properties of the system, we follow Kratsov and Malshukov [167], who consider the q_ν as external parameters of the electronic Hamiltonian, and determine the solvent configuration for which the electronic energy of the system is minimal. Using the Helmholtz-Feynman theorem we obtain the condition

$$< \partial H_t / \partial q_\nu >= 0 \tag{1.228}$$

From this, we can calculate the expectation values of the solvent coordinates in

$$< q_\nu >=(z - \sum_\sigma n_{a\sigma})g_{a\nu} \tag{1.229}$$

Substituting Eq.(1.229) into Eq.(1.225) to Eq.(1.227), we see that the slow solvent modes contribute a term

$$H_s = \sum_\sigma \left[(z - \sum_{\sigma'} < n_{a\sigma'} >) \sum_\nu \hbar\omega_\nu g_{a\nu}^2 \right] n_{a\sigma} \tag{1.230}$$

to the electronic Hamiltonian.

Consider now the interaction of the adatom with the fast modes of the interface. Normally, both the surface plasma modes of the metal and the electronic solvent modes are faster than the electron exchange between the metal and the adsorbates [168]. Then these interactions simply shift the adsorbate energy level energy ϵ_a and the correlation energy U(Hewson and Newns [169]). Estimates for the shifted values $\tilde{\epsilon}_a$ and $\tilde{U}$ are given below.

The model Hamiltonian now has is of the standard form of the Anderson model and we can use the standard theory. Due to the interaction with the metal, the adsorbate level has a width

$$\Delta(\omega) = \pi \sum_\lambda |V_{a\lambda}|^2 \delta(\omega - \epsilon_{\lambda\sigma}) \tag{1.231}$$

The width Δ is related to the self energy Σ, which is defined by

$$\Sigma(\omega) = \mathcal{P} \sum_\lambda \frac{|V_{a\lambda}|^2}{\omega - \epsilon_\lambda} \tag{1.232}$$

where $\mathcal{P}$ denotes the principal part of the sum, which excludes the term with

$$\omega = \epsilon_\lambda.$$

The effective energy of the adsorbate is then

$$E_a = \tilde{\epsilon}_a + \tilde{U} < n > +2(z - 2 < n >)E_s + \Sigma(\omega) \tag{1.233}$$

where $< n >$ is the average occupation number of the adsorbate molecule electronic level, and the factor 2 stems from the sum over the two spin states, which are assumed to be equally occupied. The solvent energy is, from Eq.(1.230)

$$E_s = (1/2) \sum_\nu \hbar\omega_\nu g_{a\nu}^2 \tag{1.234}$$

which is the energy of the solvent reorganization in Marcus theory.

| System | $\Delta = 0.5V$ | | | $\Delta = 1.0V$ | | | |
	Vac	$\alpha = 1/3$	$\alpha = 2/3$	Vac	$\alpha = 1/3$	$\alpha = 2/3$	γ_N
K^+/Hg	0.15	0.12	0.08	0.30	0.25	0.16	0.12
Rb^+/Hg	0.15	0.12	0.08	0.29	0.24	0.16	0.15
Cs^+/Hg	0.14	0.11	0.07	0.27	0.23	0.16	0.18
Cl^-/Hg	-0.21	-0.17	-0.10	-0.32	-0.27	-0.19	0.2
Br^-/Hg	-0.26	-0.20	-0.12	-0.38	-0.32	-0.22	0.34
I^-/Hg	-0.34	-0.38	-0.17	-0.47	-0.41	-0.30	0.45
Cu^{++}/Pt	1.85	1.42	1.09	1.90	1.77	1.16	1.80
Ag^+/Au	0.73	0.47	0.12	0.80	0.68	0.26	0.6
Tl^+/Pt	0.26	0.19	0.10	0.42	0.35	0.21	0.85
Pb^{++}/Ag	1.81	1.83	1.70	1.69	1.71	1.45	2.0

Table 1.3: Calculated values for the partial charge transfer,$\lambda = z - z_a$,i.e., the charge number z of the ion minus that of the adsorbate z_a[31].

Assuming that the energy width, Δ , is constant, the occupation probability $< n >$ is given by

$$< n >= (1/\pi) \int f(\omega)\frac{\Delta}{(\omega - E_a)^2 + \Delta^2} \simeq (1/2)\cot^{-1}\left[\frac{E_a - E_F}{\Delta}\right] \tag{1.235}$$

where $f(\omega)$ is the Fermi-Dirac distribution, E_F the Fermi energy of the metal, andin Eq.(1.235), the Fermi-Dirac distribution was replaced by a step function which implies $\Delta >> k_B T$, which is generally true. From $< n >$ we can estimate the partial

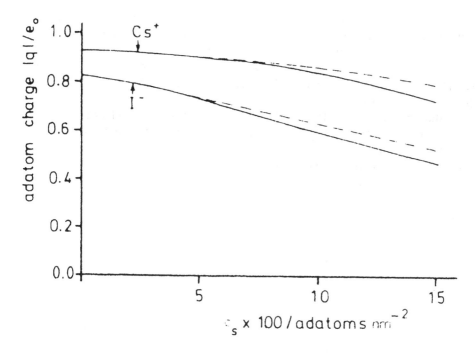

Figure 1.5: Charge on Cs$^+$ and I$^-$ ions adsorbed on mercury as a function of the adatom surface concentration in the absence of a supporting electrolyte. Plain line: cut-out disc model; Dashed line: square lattice model. The following parameters were used: l(a)=0.02nm, l=0.03nm.

charge transfer λ in Eq.(1.214)

The Anderson-Newns model yields a intuitive first principles picture of the charge transfer process in terms of parameters that need to be evaluated from some specific models. We reproduce some of the values of the charge transfer coefficients in the following table.

The calculations were performed near the potential of zero charge, when the outer potential of the metal and solution are equal. The charge transfer coefficients were calculated for energy widths of $\Delta = 0.5eV$ and$\Delta = -1eV$, which is a reasonable range for weak adsorption. The degrees of adatom adsorption were set at $\alpha = 1/3$ and $\alpha = 2/3$. $\alpha = 0$, which corresponds to the vacuum is also shown. Experimental values of the electrosorption valencies γ_N are shown in the last column of the table.

From the table we see that both the alkali ions on mercury, and the halides on mercury keep most of their charges when they are adsorbed, a fact which is in agreement with the recent in-situ experiments [170].

The transition metals such as Cu, Ag and Pb are neutralized to a large extent, also in agreement with the near edge EXAFS experiments. The values for Tl however are much lower than expected.

The adatoms adsorbed at the interface interact amongst themselves and modify the adatom-adatom interactions. They modify also the amount of charge transferred to the metal substrate.

To illustrate this point we show calculations performed for Cs^+ and I^- on mercury [171] figures 1.5 and 1.6.

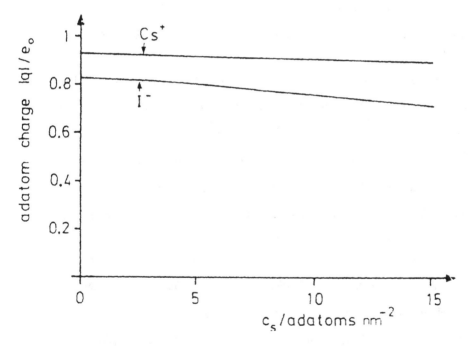

Figure 1.6: Charge on Cs^+ and I^- ions in the presence of a supporting electrolyte with a Debye length of $\kappa^{-1} = 0.96$ nm; cut-out model.

Bibliography

[1] See, for example H. D. Abruña, *Electrochemical Interfaces, Modern Techniques for In Situ Interface Characterization*, VCH Publishers, Weinheim, 1991.

[2] G. Binnig, H. Rohrer, C. Gerber and E. Weibel *Phys. Rev. Letters* 49 (1982) 57.

[3] G. Binnig, C.F. Quate and C. Gerber, *Phys. Rev. Letters,* 12 (1986) 930.

[4] R. Sonnenfeld and P.K. Hansma *Science* 232 (1986) 211.

[5] H-Y. Liu, F-R.R. Fan, C.W. Lin and A.J. Bard *J. Am. Chem. Soc.* 108 (1986) 3838.

[6] T. Hachiya, H. Honbo and K. Itaya *J. Electroanal. Chem.* 315 (1991) 275.

[7] N. Kimizuka and K. Itaya *Faraday Discuss. Chem.* 94 (1992) 275.

[8] K. Sashikata, N. Furuya and K. Itaya *J. Electroanal. Chem.* 316 (1991) 361.

[9] A. Hamelin, X. Gao and M.J. Weaver *J. Electroanal. Chem.* 323 (1992) 361.

[10] X. Gao and M.J. Weaver *J. Am. Chem. Soc.* 114 (1992) 8544.

[11] C-H. Chen, S. M. Vesecky and A.A. Gewirth *J. Am. Chem. Soc.* 114 (1992) 451.

[12] C-H. Chen and A.A. Gewirth *J. Am. Chem. Soc.* 114 (1992) 5439.

[13] X. Gao, A. Hamelin and M.J. Weaver *J. Chem. Phys.* 95 (1991) 6993.

[14] H. Honbo and K. Itaya *J. Chim. Phys.* 88 (1991) 1477.

[15] M. P. Green and K.J. Hanson *Surf. Sci. Letters* 259 (1991) L743.

[16] T. P. Russell *Mater. Sci. Reports* 5 (1990) 171.

[17] G. Materlik, M. Schmah, J. Zegenhagen and W. Uelhoff *Ber. Bunsenges. Chem.* 91 (1987) 292.

[18] J.H. White, M.J. Albarelli, H.D. Abruña, L. Blum, O.R. Melroy, M.G. Samant, G.L. Borges and J.G. Gordon II *J. Phys. Chem.* 92 (1988) 4432.

[19] O.R. Melroy, M.G. Samant, G.L. Borges, J.G. Gordon II, L. Blum, J.H. White, M.J. Albarelli, M. McMillan and H.D. Abruña *Langmuir* 4 (1988) 728.

[20] L. Blum, H.D. Abruña, J.H. White, M.J. Albarelli, J.G. Gordon, G.L. Borges, M.G. Samant and O. R. Melroy *J. Chem. Phys.* 85 (1986) 6732 ; M.G. Samant, G.L. Borges, J.G. Gordon, O.R. Melroy and L. Blum *J. Am. Chem. Soc.* 109 (1987) 5970.

[21] D.C. Koningsberger and R. Prins Editors, *X-ray Absorption: Principles, Application Techniques of EXAFS, SEXAFS and XANES*, J. Wiley, N. York, 1988.

[22] R. Feidenhans'l *Surf. Sci. Reports* 10 (1989) 105.

[23] M.G. Samant, G.L. Borges, J.G. Gordon II, O.R. Melroy and L. Blum *J. Am. Chem. Soc.* 109 (1987) 5970.

[24] M.G. Samant, M.F. Toney, G.L. Borges, L. Blum and O.R. Melroy *Surface Science Letters* 193 (1988) L29.

[25] M.G. Samant, M.F. Toney, G.L. Borges, L. Blum and O.R. Melroy *J. Phys. Chem.* 92 (1988) 220.

[26] O.R. Melroy, M.F. Toney, G.L. Borges, M.G. Samant, J.B. Kortright, P.N. Ross and L. Blum *J. Electroanal. Chem* 258 (1989) 403.

[27] G. Sauerbrey *Z. Phys.* 155 (1959) 206.

[28] D.I. Jeanmaire and R.P. Van Duyne *J. Electroanal. Chem.* 84 (1977) 1.

[29] J.N. Israelachvili *Intermolecular and Surface forces* Academic Press, San-Diego, 1992.

[30] A. Hamelin *J. Electroanal. Chem.* 144 (1983) 365.

[31] W. Schmickler and D. Henderson *Progr. Surf. Sci.* 22 (1986) 323.

[32] L. Blum *Chem. Phys. Letters* 26 (1974) 200 ; *J. Chem. Phys.* 61 (1974) 2129 ; S.A. Adelman and J.M. Deutch *J. Chem. Phys.* 60 (1974) 3935.

[33] J.P. Badiali and J. Goodisman *J. Electroanal. Chem.* 91 (1978) 151.

[34] J.P. Badiali, J. Goodisman and M.L. Rosinberg *J. Electroanal. Chem.* 130 (1981) 31.

[35] J.P. Badiali,J. Goodisman and M.L. Rosinberg *J. Electroanal. Chem.* 143 (1983) 73.

[36] J.P. Badiali, M. L. Rosinberg and J. Goodisman *J. Electroanal. Chem.* 150 (1983) 25.

[37] J.P. Badiali, M. L. Rosinberg, F. Vericat and L. Blum *J. Electroanal. Chem.* 158 (1983) 253.

[38] W. Schmickler and D. Henderson *J. Chem. Phys.* 80 (1984) 3381.

[39] D.L. Price and J.W. Halley *J. Electroanal. Chem.* 150 (1983) 347.

[40] D.L. Price and J.W. Halley *Phys. Rev. B* 38 (1988) 9357.

[41] Ch. Gruber, J.L. Lebowitz and Ph.A. Martin *J. Chem. Phys.* 75 (1981) 944.

[42] L. Blum, Ch. Gruber, J.L. Lebowitz and Ph.A. Martin *Phys. Rev. Letters* 48 (1982) 1769.

[43] L. Blum, Ch. Gruber, J.L. Lebowitz and Ph.A. Martin *J. Chem. Phys.* 78 (1983) 3195.

[44] B. Jancovici *Phys. Rev. Lett.* 46 (1981) 386.

[45] B. Jancovici *J. Physique Lett.* 42 (1981) L226.

[46] A. Alastuey and B. Jancovici *J. Physique* 42 (1981) 1.

[47] B. Jancovici *J. Stat. Phys.* 28 (1982) 43.

[48] B. Jancovici *J. Stat. Phys.* 29 (1982) 263.

[49] B. Jancovici *J. Stat. Phys.* 34 (1984) 803.

[50] E.H. Lieb and J.L. Lebowitz *Adv. in Math.* 9 (1972) 316 ; J.L. Lebowitz and Ph. A. Martin *Phys. Rev. Lett.* 48 (1982) 1769.

[51] A. Alastuey and Ph.A. Martin *Europhys. Letters* 6 (1988) 385.

[52] F.H. Stillinger and R. Lovett *J. Chem. Phys.* 48 (1968) 3858 ; *ibid* 49 (1968) 1991.

[53] C.W. Outhwaite *Chem. Phys. Letters* 24 (1973) 73.

[54] S.L. Carnie and D.Y.C. Chan *Chem. Phys. Letters* 77 (1981) 437.

[55] L. Blum *Advances in Chemical Physics*, S. A. Rice and I. Prigogine, Editors, J. Wiley, New York, 78 (1991) 171.

[56] D.A. Huckaby and L. Blum *J. Chem. Phys.* 92 (1990) 2646.

[57] L. Blum and D.A. Huckaby *J. Chem. Phys.* 94 (1991) 6887.

[58] D. A. Huckaby and L. Blum *J. Electroanal. Chem.* 315 (1991) 255.

[59] R. J. Baxter *Exactly Solved Models in Statistical Mechanics*, Academic Press, New York, 1982.

[60] J. D. Jackson *Classical Electrodynamics*, J.Wiley, New York, 1962

[61] M.F. Toney, J.N. Howard, J. Richer, G.L. Borges, J.G. Gordon II, O. R. Melroy, D. Yee and L.B. Sorensen *Phys. Rev. Lett.* 75 (1995) 4472.

[62] G. Gouy *J. Phys.* 9 (1910) 457.

[63] D.L. Chapman *Phil. Mag.* 25 (1913) 475.

[64] S. Levine *Proc. Phys. Soc.* 64 (1951) 781 ; *ibid.* 66 (1951) 357.

[65] J.L. Lebowitz and J.K. Percus *J. Math. Phys.* 4 (1963) 116; *ibid.* 4 (1963) 248.

[66] J.P. Hansen and I.R. McDonald *Theory of Simple Liquids,* Academic Press, New York, 1976.

[67] L. Blum and G. Stell *J. Stat. Phys.* 15 (1976) 439.

[68] D.E. Sullivan and G. Stell *J. Chem. Phys.* 67 (1977) 2567.

[69] M. Born and H.S. Green *Proc. Roy. Soc. London* 188 (1988) 10.

[70] J. Yvon *La théorie statistique des fluides et l'équation d'état*, Actualités scientifiques et industrielles 203, Paris, 1935.

[71] G.M. Torrie and J.P. Valleau *J. Chem. Phys* 73 (1980) 5807.

[72] J.P. Valleau and G.M. Torrie *J. Chem. Phys.* 76 (1982) 4623.

[73] C. Caccamo, G. Pizzimenti and L. Blum *J. Chem. Phys.* 84 (1986) 3327.

[74] T.L. Croxton and D.A. McQuarrie *Mol. Phys.* 42 (1981) 141.

[75] M. Plischke and D Henderson *J. Chem. Phys.* 90 (1989) 5738.

[76] D. Henderson and M. Plischke *J. Phys. Chem.* 92 (1988) 7177.

[77] P.J. Colmenares and W. Olivares *J. Chem. Phys.* 90 (1986) 1977.

[78] P.J. Colmenares and W. Olivares *J. Chem. Phys.* 88 (1988) 3221.

[79] D.J. Henderson F.F. Abraham and J.A. Barker *Mol. Phys.* 31 (1976) 1291.

[80] J.K. Percus *J. Stat. Phys.* 15 (1976) 423.

[81] D.J. Henderson, L. Blum and W.R. Smith *Chem. Phys. Letters* 63 (1979) 381.

[82] M. Lozada-Cassou, R. Saavedra-Barrera and D.J. Henderson *J. Chem. Phys.* 77 (1982) 5150.

[83] M. Lozada-Cassou and D.J. Henderson *J. Phys. Chem.* 87 (1983) 2821.

[84] S.L. Carnie and D.Y.C. Chan *Chem. Phys. Letters* 77 (1981) 437.

[85] S.L. Carnie and D.Y.C. Chan *Mol. Phys.* 51 (1984) 1047.

[86] S.L. Carnie and D.Y.C. Chan *J. Chem. Phys.* 73 (1980) 2949.

[87] S.L. Carnie and G.M. Torrie *Adv. Chem. Phys* 56 (1984) 141.

[88] P. Ballone, G. Pastore and M. Tosi *J. Chem. Phys.* 81 (1984) 3174.

[89] Y. Rosenfeld and L. Blum *J. Chem. Phys.* 85 (1986) 2197.

[90] V.M.B. Marconi, J. Wichen and F. Forstmann *Chem. Phys. Letters* 107 (1984) 609

[91] P. Nielaba and F. Forstmann *Chem. Phys. Letters* 117 (1985) 46.

[92] T. Alts, P. Nielaba, B. D'Aguanno and F. Forstmann *Chem. Phys.* 111 (1987) 223.

[93] P. Nielaba, T. Alts, B. D'Aguanno and F. Forstmann *Phys. Rev. A* 34 (1986) 1505.

[94] B. D'Aguanno, P. Nielaba, T. Alts and F. Forstmann *Chem. Phys.* 85 (1986) 3476.

[95] R. Kjellander and S. Marcelja *Chem. Phys. Letters* 112 (1984) 49.

[96] R. Kjellander and S. Marcelja *J. Chem. Phys.* 82 (1985) 2122.

[97] R. Kjellander and S. Marcelja *Chem. Phys. Letters* 127 (1986) 402.

[98] Y. Rosenfeld *Phys. Rev. Lett.* 63 (1989) 980.

[99] Y. Rosenfeld *Phys. Rev. A* 42 (1990) 5978.

[100] Y. Rosenfeld *J. Chem. Phys.* 93 (1990) 4305.

[101] Y. Rosenfeld, D. Levesque and J.J. Weis *J. Chem. Phys.* 92 (1990) 6918.

[102] L. Blum and Y. Rosenfeld *J. Stat. Phys.* 63 (1991) 1177.

[103] E. Kierlik and M. L. Rosinberg *Phys. Rev. A* 44 (1991) 5025.

[104] R. Evans and T. Sluckin *Mol. Phys.* 40 (1980) 413; *J. Chem. Soc. Faraday II* 77 (1981) 575.

[105] J.P. Badiali *J. Electroanal. Chem.* 31 (1986) 149.

[106] *CRC Handbook of Chemistry and Physics*, R.C. West and M.J. Astle, Editors, Cleveland, 1983.

[107] S.L. Carnie and D.Y.C. Chan *J. Chem. Phys.* 73 (1980) 2949.

[108] L. Blum and D. Henderson *J. Chem. Phys.* 74 (1981) 1902.

[109] W.R Smith and D. Henderson *J. Stat. Phys.* 19 (1978) 191.

[110] L. Blum, D. Henderson and R. Parsons *J. Electroanal. Chem.* 161 (1984) 389.

[111] J.S. Perkyns, P.H. Fries and G.N. Patey *Mol. Phys.* 57 (1986) 529.

[112] R.J. Baxter *J. Chem. Phys.* 52 (1970) 4559.

[113] E. Armand and M.L. Rosinberg *J. Electroanal. Chem.* 302 (1991) 191.

[114] R.B. Potts *Phys. Rev.* 88 (1952) 352.

[115] A.J. Bard and L. R. Faulkner *Electrochemical Methods* J. Wiley, New York, 1980.

[116] G.V. Hevesy *Physik Z.* 13 (1912) 715.

[117] K.F. Herzfeld *Physik Z.* 14 (1914) 29.

[118] J.W. Schultze and K.J. Vetter *J. Electroanal. Chem.* 44 (1973) 63.

[119] L. Blum *J. Stat. Phys.* 75 (1994) 971.

[120] D. M. Kolb, K. Al Jaaf-Golze and M. S. Zei *DECHEMA Monographien Verlag Chemie Weinheim,* 12 (1986) 53 ; M. Zei, G. Qiao, G. Lehmpful and D. M. Kolb *Ber. Bunsenges. Phys. Chem.* 91 (1987) 3494.

[121] O. M. Magnussen, J. Hotlos, R.J. Nichols, D.M. Kolb and R.J. Behm *Phys. Rev. Letters* 64 (1990) 2929.

[122] M.G. Samant, K. Kunimatsu, H. Seki and M. R. Philpott *J. Electroanal. Chem.* 280 (1990) 391.

[123] L. A. Barnes, B. Liu and M. R. Philpott, Proceedings of the symposium *Microscopic Models of Electrode-Electrolyte Interfaces,* J. Woods Halley and L. Blum, Editors *Proc. Electrochemical Soc.* 93-5 (1993) 272.

[124] L. Blum and D.A. Huckaby *J. Electroanal. Chem.* 375 (1994) 69.

[125] P.A. Rikvold, J.B. Collins, G.D. Hansen and J.D. Gunton *Surf. Sci.* 203 (1988) 500.

[126] J.B. Collins, P. Sacramento, P.A. Rikvold and J.D. Gunton *Surf. Sci.* 221 (1989) 277.

[127] J.B. Collins, P.A. Rikvold and E.T. Gawlinski *Phys. Rev. B* 38 (1988) 6741.

[128] D.M. Kolb in *Advances in Electrochemistry and Electrochemical Engineering* 11 125, H. Gerischer and C. W. Tobias, Editors, J. Wiley, New York, 1978.

[129] L. Blum and D.A. Huckaby, Proceedings of the symposium *Microscopic Models of Electrode-Electrolyte Interfaces,* J. Woods Halley and L. Blum, Editors *Proc. Electrochemical Soc.* 93-5 (1993) 232.

[130] P.A. Rikvold *Electrochim. Acta* 36 (1990) 171.

[131] A. Bewick and B. Thomas *J. Electroanal. Chem.* 85 (1977) 329.

[132] C. Buess-Hermann in *Trends in Interfacial Electrochemistry*, A Fernando Silva, Ed. Reidel Dordrecht, Holland, 1984.

[133] A.N. Frumkin *Ann. Physik* 35 (1926) 792.

[134] E.H. Stanley, *Introduction to Phase Transitions and Critical Phenomena*, Oxford University Press, London, 1971.

[135] J. Goodisman *Electrochemistry: Theoretical Foundations*, J. Wiley and Sons, New York, 1987.

[136] T. Hachiya, H. Honbo and K. Itaya *J. Electroanal. Chem.* 315 (1991) 275.

[137] G. Tourillon, D. Guay and A. Tadjeddine *J. Electroanal. Chem.* 289 (1990) 263.

[138] A. Tadjeddine, D. Guay, M. Ladouceur and G. Tourillon *Phys. Rev. Letters* 66 (1991) 2235.

[139] D.M. Kolb *Ber. Bunsenges. Phys. Chem.* 92 (1988) 1175.

[140] C.H. Chen, S.M. Vesecky and A.A. Gewirth *J. Am. Chem. Soc.* 114 (1992) 5439.

[141] M. Hölzle *PhD Thesis*, Ulm, 1995.

[142] I.H. Omar, H.J. Pauling and K. Jüttner *J. Electr. Soc.* 140 (1993) 2187.

[143] D.M. Kolb, Schering Lecture (1991).

[144] G.S. Joyce *Phil. Trans. Roy. Soc. London A* 325 (1988) 643.

[145] P. Zelenay, L. M. Rice and A. Wieckowski *Surf. Sci.* 256 (1991) 253.

[146] L.M. Rice-Jackson *PhD Thesis* Univ. of Illinois, Urbana, 1990.

[147] S. Trasatti *J. Electroanal. Chem.* 33 (1971) 351.

[148] S. Naya *Prog. Theor. Phys.* 11 (1954) 53.

[149] D.A. Huckaby and L. Blum Proceedings of the symposium *X-Ray Methods in Corrosion and Interfacial Electrochemistry*, A. Davenport and J.G. Gordon, Editors *Proc. Electrochemical Soc.* 92-1 (1992) 139.

[150] B. Kurtyka, M. Kaisheva and R. De Levie *J. Electroanal. Chem.* 341 (1992) 343 ; R. De Levie and A. Vogt *J. Electroanal. Chem.* 341 (1992) 353.

[151] F. Lantelme and E. Cherrat *J. Electroanal. Chem.* 297 (1991) 409.

[152] P. Hohenberg and W. Kohn *Phys. Rev. B* 136 (1964) 864.

[153] L. Blum, M. Legault and P. Turq *J. Electroanal. Chem.* 379 (1994) 35.

[154] W. Kohn and L.J. Sham *Phys. Rev. A* 140 (1965) 1133.

[155] N.D. Lang and W. Kohn *Phys. Rev. B* 1 (1970) 4555 ; *ibid* 3 (1971) 1215; *ibid* 7 (1973) 2541.

[156] J. Friedel *Adv. Phys.* 3 (1954) 446.

[157] S. Lundquist and N.H. March, Eds. *Theory of the Inhomogeneous Electron Gas,* Plenum Press, 1983.

[158] J. Vanneminus and H.F. Budd *Solid State Comm.* 15 (1974) 1739; H.F. Budd and J. Vanneminus *Phys. Rev. Lett.* 31 (1973) 1212; *ibid* 31 (1973) 1430E.

[159] W.Lorenz and G. Salie *Z. Phys. Chem.* NF29 390 (1961) 408.

[160] K.J. Vetter and J.W. Schultze *Ber. Bunsen Ges. Chem.* 76 920 (1973) 927; J.W. Schultze and F.D. Koppitz *Electrochimica Acta* 21 (1977) 81; J.W. Schultze and K.J. Vetter *J. Electroanal. Chem.* 44 (1973) 63.

[161] J.E.B. Randles *Trans Farad. Soc.* 52 (1956) 1573.

[162] J.P. Muscat and D.M. Newns *Progr. Surf. Sci.* 9 (1978) 1.

[163] J.P. Gadzuk in *Surface Phys. of Materials,* J. M. Blakely Ed , Academic Press, New York, 1975.

[164] W. Schmickler *J. Electroanal. Chem.* 100 (1979) 533.

[165] A.A. Kornyshev and W. Schmickler *J. Electroanal. Chem.* 185 (1985) 253.

[166] J. Ulstrup *Charge Transfer Processes in Condensed Media,* Springer, Berlin, 1979.

[167] V.E. Kratsov and A.G. Malshukov *Soviet Phys. JETP* 48 (1978) 248.

[168] J. P. Gadzuk, J.K. Hartmann and T.N. Rodin *Phys. Rev. B* 4 (1971) 241.

[169] A. C. Hewson and D. M. Newns *J. App. Phys. Suppl* 2 (1974) 2121.

[170] L. Blum *Advances in Chem. Phys.* 78 (1991) 171.

[171] A.A. Kornyshev and W. Scmickler *J. Electroanal. Chem.* 20 (1986) 21.

Chapter 2

Ions at liquid/air and liquid/liquid interfaces

Interfacial properties of electrolyte solutions have been studied experimentally and theoretically. In this chapter, we first present basic notions about capillarity . Then we consider the theoretical models available and, last, we list the main experimental results reported in the literature concerning liquid/air and liquid/liquid interfaces.

2.1 Capillarity: basic concepts

The aim of this section is to present basic concepts on capillarity phenomena. The reader is referred to other books [1, 2] for extensive accounts of the subject.

2.1.1 Surface tension

Everyone has played with soap bubbles. And everyone has noticed that one has to *blow* air into the bubble so as to form it. Moreover, if one creates two bubbles and connects them with a small tube, then the smaller of the two bubbles shrinks while the other enlarges (Figure 2.1); at the end of the process the small bubble just covers the mouth of the tube and its radius is equal to that of the large bubble.

These simple experiments show that the air pressure inside the bubble is larger than atmospheric pressure, by an amount Δp and that the latter increases when the bubble radius r decreases.

Thus, work is necessary to increase the area of a liquid film. The physical origin of this phenomenon is at the microscopic level: work is required to bring molecules

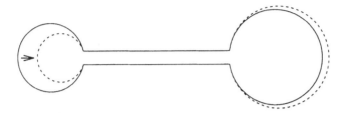

Figure 2.1: The smaller bubble shrinks until its radius is equal to that of the larger bubble.

from the inside to the surface where the molecules are less bonded to the other solvent molecules.

Mathematically the work necessary to increase the interfacial area by dA can be expressed as

$$\delta W = \gamma \, dA \qquad (2.1)$$

where γ denotes the surface tension. The SI unit for γ is Joules per square meter (J m^{-2})or Newton per meter (N m^{-1}). In the literature many values can be found in the cgs unit system: surface tensions given in dyn cm^{-1} have the same numerical value as those given in mN m^{-1}.

Eq 2.1 is valid for liquid/air, liquid/solid and liquid/liquid interfaces.

For a soap bubble in equilibrium, eq 2.1 leads to

$$\Delta p = 4\gamma/r \qquad (2.2)$$

if the film is not too thin and bearing in mind that the liquid film has two liquid/air interfaces.

2.1.2   The Gibbs dividing plane

A liquid surface is the location of violent agitation on the molecular scale with molecules passing rapidly back and forth between the interfacial zone and the bulk phase(s). If we consider the concentration of one of the phases, denoted arbitrary by 1, we will find that this quantity varies continuously in space from a value $C_1^{(1)}$ in the region where it is the major component to a value $C_1^{(2)}$ in the other region. This is illustrated in Figure 2.2.

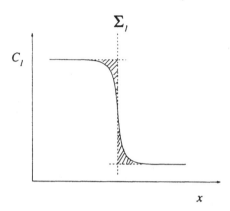

Figure 2.2: Variation of density in the interfacial region.

The interfacial zone is therefore a blurred region for which mention of a definite "interface" requires a suitable definition. This can be achieved by reference to the so-called Gibbs procedure: it is simply defined in Figure 2.2 in which the vertical line Σ_1 is such that the two hatched zones have equal areas.

An interesting consequence of this definition is that it can be used to define univocally surface excess amounts. As shown in Figure 2.3, where the case of an amphiphilic compound adsorbed at a liquid/liquid interface is considered, it can be defined as the algebraic sum of the hatched zones. Classically this quantity is denoted by $\Gamma_i^{(1)}$ for a compound i, with reference to solvent 1. The unit for Γ is moles per square meter (mol m^{-2}).

As a consequence, we find that the definition of the Gibbs dividing plane Σ_1 entails

$$\Gamma_1^{(1)} \equiv 0 \qquad (2.3)$$

Since the phase chosen as a reference generally determines the value for Γ_i, it should be clearly mentioned in each study aimed at assessing surface excess concentrations.

2.1.3 Thermodynamics: the Gibbs equation

The Gibbs convention presented in the previous section can be applied to extensive thermodynamic functions. So, the total internal energy of a system composed of

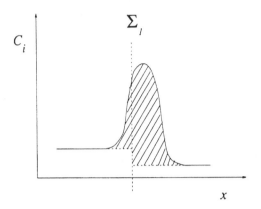

Figure 2.3: Definition of the Gibbs surface Σ_1.

two phases 1 and 2 in contact may be written as

$$U_t = U_1 + U_2 + U_\sigma \tag{2.4}$$

in which U_σ represents the excess surface energy with the choice of the Gibbs dividing plane as given above (for instance with respect to phase 1).

For an homogeneous multicomponent bulk phase one has that

$$U = TS - pV + \sum_i \mu_i n_i \tag{2.5}$$

with T the temperature, S the entropy, p the pressure, V the volume, μ_i the chemical potential of species i and n_i the number of moles.

By using eqs 2.1, 2.4 and 2.5 for each phase, it is easy to show that a similar relation exists for U_σ in which γA replaces $-pV$, that is

$$U_\sigma = TS_\sigma + \sum_i \mu_i n_{i,\sigma} + \gamma A \tag{2.6}$$

in which $n_{i,\sigma}$ is the excess surface number of moles of i.

Then, the Gibbs-Duhem relation that expresses the relation between the intensive surface excess thermodynamic variables reads

$$S_\sigma dT + \sum_i n_{i,\sigma} d\mu_i + A d\gamma = 0 \tag{2.7}$$

For a small change at constant temperature we get from eq 2.7

$$d\gamma = - \sum_{i} \Gamma_i d\mu_i \qquad (2.8)$$

For a two-component system eq 2.8 further reduces to

$$\Gamma_2^{(1)} = - \left(\frac{\partial \gamma}{\partial \mu_2} \right)_T \qquad (2.9)$$

with the convention given by eq 2.3. In this equation μ_2 is the chemical potential of the solute; it can be estimated experimentally by various methods.

Eqs 2.8 constitutes the Gibbs equation that is of uppermost importance in surface chemistry. It can be used to derive surface excess amounts from experimental surface tension studies.

2.2 Theoretical descriptions

The first theoretical work was proposed by Wagner [3] in 1924 in the early days of the Debye-Hückel (DH) theory. Subsequently in 1934, on the basis of Wagner's work, Onsager and Samaras [4] (OS) succeeded in deriving analytical expressions. The OS calculation constitutes an important benchmark description of the surface tension of a simple primitive model electrolyte solution. Further studies have then been aimed at improving the OS treatment. The latter was first reexamined in 1956 by Buff and Stillinger [5], using a virial expansion method. But it is at the end of the 70's that appeared a renewed interest in understanding the surface properties of electrolyte solutions. The theoretical tools that have been used include the Born-Green-Yvon hierarchy, the random phase approximation (RPA) and the Poisson-Boltzmann equation. All these theories have been developed at the level of the primitive model of ionic solutions, consisting of charged hard spheres in a dielectric continuum.

Besides, computer simulations, based on molecular dynamics calculations, have been performed, with a first study in 1989 [6]. In contrast with the former descriptions, these simulations were aimed at a better account of the solvent effects close to the interface.

The surface tension of Debye-Hückel electrolytes:
OS theory

The OS model was developed in a framework proposed earlier by Wagner [3], in which the ions in the vicinity of the interface are repelled from the latter by electrostatic image forces (see Figure 2.4).

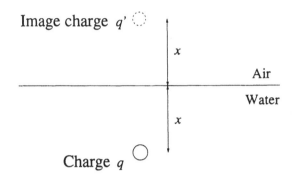

Figure 2.4: Repulsion of a charge q by its image charge.

The range of exclusion is approximately given by the Debye screening length, which expresses that the repulsion forces are screened out at this distance by the effect of the ionic atmosphere.

Because it represents an important reference in the field we give a short account of the OS theory.

The electrostatic potential ψ was assumed to obey the Debye-Hückel equation

$$\Delta\psi - \kappa^2\psi = 0 \qquad (2.10)$$

in which κ is the Debye screening parameter given by

$$\kappa^2 = 4\pi\lambda \sum_i \rho_i z_i^2 \qquad (2.11)$$

in the MKS unit system, with ρ_i the number density of species i, z_i its valence number and

$$\lambda = e^2/(4\pi\varepsilon_0\varepsilon k_B T) \qquad (2.12)$$

the Bjerrum distance (*ca.* 7 Å in water at 25°C), in which e is the unit charge, ε_0 the permittivity of a vacuum, ε the relative permittivity of the solvent, k_B the Boltzmann constant and T the temperature.

Since the ions are repelled from the interface, κ is a function of the distance from the interface. In order to simplify the problem the authors made the approximation

$$\kappa(x) \simeq \kappa(\infty) = \kappa \qquad (2.13)$$

where x is the distance from the interface and κ is the value in the bulk solution. They showed that the corresponding error was small for dilute solutions.

Then, for a charge q, the valence of its image charge is

$$q' = \frac{\varepsilon - 1}{\varepsilon + 1} q \simeq q \tag{2.14}$$

because $\varepsilon \gg 1$ for water, and the electrostatic potential at the point where the charge q is located, caused by the charge q' is

$$\psi = \frac{q}{4\pi\varepsilon_0\varepsilon} \frac{\exp(-\kappa r)}{r} \tag{2.15}$$

where $r = 2x$. Therefore, the electrostatic energy W of the charge q caused by the fictitious image charge q', or the potential of mean force on an ion near the surface, is

$$W(x) = \frac{1}{2} q\psi \tag{2.16}$$

or

$$\beta W(x) = \lambda \left(\frac{q}{e}\right)^2 \frac{\exp(-2\kappa x)}{4x} \tag{2.17}$$

with λ given by eq 2.12 and $\beta = 1/k_BT$. It has the form of the repulsive potential of the ion's own image, screened by an exponential factor typical of DH theory.

This energy was defined as the adsorption potential: it represents also the amount of work that has to be provided to bring a charge q from the interior of the solution to a point at a distance x from the interface.

Thus the concentration of either ion was expressed by the Maxwell-Boltzmann formula

$$C(x) = C \exp\left[-\beta W(x)\right] \tag{2.18}$$

with the use of eq 2.17 and with $C = C(x = \infty)$ the bulk concentration.

Then the adsorbed amount reads

$$\Gamma = \int_0^\infty [C(x) - C] \, dx \tag{2.19}$$

Upon integration and applying the Gibbs equation to a 1:1 electrolyte led to the following limiting law

$$\gamma - \gamma_0 = \frac{80.00}{\varepsilon} C \log_{10}[1.130 \times 10^{-13}(\varepsilon T)^3/C] \tag{2.20}$$

where the solute concentration C is expressed in moles per liter and γ is in mN m^{-1}, or dynes cm^{-1}. Moreover, this expression [5] included more recent values for the physical constants.

For higher concentrations, expressions were derived that accounted also for the finite size of the ions. This was done in the DH approximation, by replacing eq 2.15 by an expression that includes a mean distance of approach for the ions. It was found that the result for γ was weakly sensitive to the precise value of this distance.

This result was regarded as a confirmation of the experimental observation that "for a given concentration the increase in surface tension is the same for all uni-univalent electrolytes examined, within the limits of the experimental error". Besides, the numerical results from the model were in qualitative agreement with experimental results [7, 8] reported by Schwenker and Heydweiller: Schwenker's values, likely to be the more reliable ones, were about 20-30% in excess over the theoretical OS values.

It must be underlined here that a more recent work [9] concluded to the possibility of a different dependence of the surface tension on concentration: this effect was traced to the nonzero relative solubility of the salt between the coexisting phases, even for a vapor phase, and to the dissymmetric ion-solvent interactions in both phases. Then the surface tension was predicted to *decrease* proportional to the square root of the concentration, as $C^{1/2}$ instead of $C \ln C$, in the limiting region of low concentrations. This result was regarded as a possible justification for the fact that experimental data are sometimes empirically fitted with such a power law.

Lastly, the influence of ion polarizability has been approached recently [10].

Ion-free layer model

In this simple picture, a layer of arbitrary thickness δ at the upper limit of the liquid phase was assumed [11] to be ion-free (ions being excluded by an infinite potential barrier). The ion distribution below this layer was calculated by a method similar to OS. The formula obtained was

$$\gamma - \gamma_0 = 10^{-3} N_A kT\delta \, \nu m\phi + \frac{kT}{32\pi\delta^2} I_0(2\delta\kappa) \tag{2.21}$$

with N_A the Avogadro's number, m the molality, ν the number of ions produced by one solute molecule, ϕ the molal osmotic coefficient of the solution and I_0 an integral tabulated by the author.

The first term of the r.h.s. of this equation represents the contribution of the ion-free layer, as can be shown simply by use of the Gibbs equation for the solvent in which the surface excess of solvent is proportional to δ. The second term is the increment due the deficit of ions beyond the ion-free layer.

At very low concentrations the first term becomes negligible in comparison with the second term and the OS limiting law is recovered. Physically this is because, at very low concentrations, the surface zone is free from ions in the OS theory.

Virial expansion method

Here [5], the molecular distribution functions were calculated by linearization of a generalized form of Kirkwood's integral equation [12, 13] and the increase in surface tension was computed from the molecular theory of Buff [14, 15] for this property. The ions were taken as point charges. The singlet and pair distribution functions were evaluated for very dilute ionic solutions. Then they were introduced into the statistical mechanical formulas for surface tension.

The results of the theory were in better agreement with the experimental results of Schwenker [7] than the earlier OS calculation. So, at 0.16 mol L^{-1}, the theoretical value was smaller than the experimental value by about 10% as compared to 30% for the OS value.

Use of the Born-Green-Yvon hierarchy

The Born-Green-Yvon (BGY) integral equation approach was used [16] for dilute electrolyte solution, with a modification which ensures electroneutrality (BGY+EN). Ions were taken as hard charged spheres of the same diameter, corresponding to the closest ionic approach distance and to twice the closest approach distance to the interface (restricted pritive model, see Figure 2.5).

The results were compared with results from Monte-Carlo and hypernetted-chain (HNC) computations. It was concluded that the BGY+EN theory yielded density profiles in close agreement with Monte-Carlo computations, even for a 1 mol L^{-1} 1:1 electrolyte. In comparison with these, the OS result was found to underestimate the ion depletion and therefore the increase of surface tension caused by the ions.

In a subsequent work [17], more extensive calculations using the BGY equation were reported. Surface tensions were derived from this treatment and a comparison to experimental data was made in the range 0.1-1.5 mol L^{-1}. This theory is internally consistent, in contrast with the Buff-Stillinger theory [5] (the virial expansion method described above) which is not, probably because of the linearization approximation. Besides, this study incorporated an additional "soft" interaction of the ion

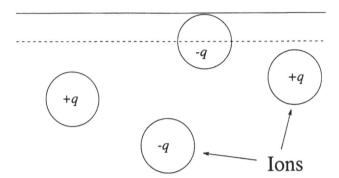

Figure 2.5: Restricted primitive model; closest approach distance to the interface (dashed line).

with the wall (the interface), which accounted reasonably for the rearrangement of solvent structure as an ion approached the interface. The corresponding potential was chosen as being proportional to the volume of hydration sheath that has to be displaced, as shown in Figure 2.6.

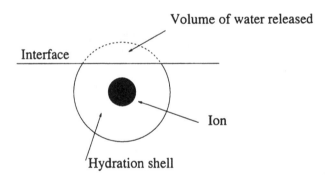

Figure 2.6: Release of solvation water as the ion approaches the interface.

A proportionality constant α was introduced as an adjustable parameter, characteristic of the binding energy per water molecule. With these assumptions the experimental values of Johansson and Eriksson [18], for KCl and KI, could be described precisely with the choice of suitable values for α.

Random phase approximation

A theory [19] was developed, in the framework of the random phase approximation (RPA), for the salt concentration profile and ion-ion correlations near surfaces of dilute electrolyte solutions. The formulation of the problem was based on the study of the cluster integral contributions: the chain diagrams for the correlation functions and the ring diagrams for the thermodynamic properties. The way in which these quantities are altered by the influence of the surface was examined. It was possible to lift completely the OS approximation that the ratio f of the image charge q' to the charge q is equal to unity. This property allows to study the thermodynamics of interfaces between coexisting phases of more comparable dielectric constants than aqueous liquid-vapor solutions: near critical surfaces of electrolyte solutions, solutions bounded by semi-conductor surfaces or ice-aqueous electrolyte solution interface. Results for $f \neq 1$ were compared with those from OS theory: here too, it was found that the OS theory leads to an underestimation of the ion depletion near the surface.

In a later work [20], the theory was extended to treat higher concentration solutions. A new analytical treatment of the RPA was obtained. The results were compared with available computer simulation data and the agreement was excellent. Besides, a compensation effect was detected in the improved treatment of the RPA, which provided an explanation for the fact that the OS profiles are closer to the computer experimental results for 1.0 mol L^{-1} than they are for 0.1 mol L^{-1}.

Modified Poisson-Boltzmann approximation

The Poisson-Boltzmann (PB) equation [21] combines the Poisson's equation for the electrostatics with the Boltzmann's equation for the thermodynamic fluctuations. It yields the DH equation upon linearization of the exponential terms.

The modified Poisson-Boltzmann (MPB) approximation is the product of many theoretical investigations into improvements of the PB approach of Gouy and Chapman based on classical statistical mechanics [22, 23] for the description of the electric double layer at a charged plane interface. These extensions originated in a work of Kirkwood [24] who used the Güntelberg charging process to examine the fluctuation terms neglected in the DH theory of electrolytes.

The MPB approximation has been used recently, in 1991, by Bhuiyan *et al.* [25] for the surface tension problem. The excess surface tension was determined by numerical integration of the Gibbs adsorption isotherm (the Gibbs equation), with the electrolyte activity obtained from the bulk MPB approximation. A simple model of the interface, used previously in another study [16], was adopted. In an earlier study [26], this version of the MPB equation had been reported to be thermodynamically consistent to a very good approximation and of comparable accuracy to the HNC theory. Here too, the ions experience volume exclusion forces: interionic repulsion for $r < \sigma$ and exclusion from the interfacial region for $x < \sigma/2$, with x the distance from the wall and σ the common ionic diameter of both ions (restricted case). The case of symmetric $z : z$ electrolytes was considered.

The calculated wall/ion density profiles were compared with the corresponding Monte-Carlo data for 1:1 and 2:2 valency type electrolytes, and the results for the excess surface tension were compared with the experimental values for LiCl, NaCl, KCl, NH_4Cl as 1:1 electrolytes and $MgSO_4$ and $ZnSO_4$ as 2:2 electrolytes up to concentrations of 1.8 mol L^{-1}. Good agreement with the MC simulations were observed at low concentration, typically 0.1 mol L^{-1}; however, with increasing concentration (up to 1 mol L^{-1}) the deviations became more pronounced but were generally smaller than those observed with the RPA approximation [20]. The MPB distributions for the divalent ions revealed much stronger image repulsion from the wall while the thickness of the desorption layer was reduced due to enhanced screening of image interactions.

The MPB computed surface tension for 1:1 electrolytes yielded a limiting behaviour that was similar to those from RPA and OS theories. However, at finite concentration, the MPB approximation predicted excess surface tensions that were appreciably bigger than than that found by the other two theories. A set of MPB calculations was carried out using the mean ionic diameters fitting the DH theory to the bulk thermodynamic properties in experimental systems. The agreement with the experimental data was remarkably good for all the systems studied. In the case of the two 2:2 salts considered the agreement was excellent with a value for the closest approach distance that was somewhat smaller (5 Å) than the DH value (5.45 Å); in this case the surface tension increment was greater than 4 mN m^{-1} at the highest concentration of *ca.* 1.8 mol L^{-1}.

Computer experiments

Here computer experiments refer exclusively to equilibrium molecular dynamics (MD) calculations.

MD simulations of the structure of water at the interface between pure water and its vapor have shown [27]-[31] that the molecular dipoles of water tend to lie parallel to the surface, though with a net dipole moment of the interfacial layer which points into the liquid and persists several molecular layers into the bulk. Similar preferred orientations have been observed experimentally [32]. Besides, MD calculations have been used [33] to interpret [34] (electrostatic) surface potential measurements.

The first MD study including a solute was done by Wilson, Pohorille and Pratt in 1989, about the interactions of a sodium ion with a water liquid-vapor interface at 320 K. They presented results for the density profile of Na^+ in a lamella of liquid water, ca. seven molecular layers thick (a compromise between bulk and interfacial scales), in coexistence with its vapor. The model taken for the water was similar to those in previous calculations on the liquid-vapor interface of pure water [28, 29]: the so-called TIP4P pair potential model was adopted for intermolecular forces between water molecules and a corresponding potential was used for the interaction between the sodium ion and the water [35, 36]. The system consisted of one sodium ion immersed in a assembly of 342 water molecules. The liquid-vapor interface was created by expanding one edge of a cubic simulation box from 21.7 Å to 65.1 Å. The main results were as follows. The water density profiles were the same as for pure water. The ion density profile was found to disagree with the profile derived from a simple dielectric continuum model, with the latter extending too close to the interface. The data suggested that dielectric models should locate the dielectric surface (at which the model profile may match the MD data) at least two molecular diameters inside the liquid phase. However, appreciable ion concentration can be anticipated beyond this dielectric surface, near the interface. The quantity

$$-kT \ln[\rho_{ion}(z)/\rho_{ion}(z')] \qquad (2.22)$$

may be interpreted as the free energy cost, or minimum work, required to translate a sodium ion from a depth z' to z, ρ_{ion} being the ion number density. Therefrom, the work required to push an ion from the interface toward the inside of the lamella was found to be significantly smaller when calculated from MD than from the model.

This work was extended later [37] to the case of anions F^- and Cl^- in presence of the Na^+ ion. Here too, the system consisted of 342 water molecules and 1 ion, either Na^+, F^- or Cl^-. Periodic boundary conditions were applied in all directions and the TIP4P model was used. As suggested by experimental work in which surface potentials were measured [38], an ionic double layer was found where anions penetrate closer to the interface than do cations. Correspondingly less free energy is needed to move the anions Cl^- and F^- to the interface than to move the cation Na^+. This difference arises from the asymmetric orientational distribution of water molecules at the pure water liquid-vapor interface, mentioned at the beginning of this section: the interaction of the interfacial water molecules with the anions is more favorable than with Na^+ because the polarization which develops around the anions perturbs the interfacial water structure much less than does the polarization around the cation. The free energy curves were compared with predictions of simple dielectric models and, again, these models were shown to give a poor description.

Moreover, the ions were found to retain their first hydration shell at the interface; the anions retain also part of their second shell, while Na^+ does not. Besides, the mobilities of the ions were estimated: they led to the observation of an increase in mobility (by factors greater than 2), particularly for Na^+; this result was in satisfying agreement with the prediction of a simple hydrodynamic model [39], in which the quasisteady Stokes equation was solved for a sphere approaching a surface capable of slight deformation.

Similar conclusions have been reported at the same time in another work [40].

MD simulations have also been developed to understand some equilibrium properties of liquid/liquid interfaces.

The interface between two nonpolar atomic solvents interacting through Lennard-Jones-type potentials has been studied [41]: diffusion' in the interfacial region was found to be anisotropic.

MD calculations have been used to simulate the water/1,2-dichloroethane (D-CE) interface [42]. The most important result was that a molecularly sharp interface was found, with capillary wavelike distortions whose structure and dynamics closely resembled those expected from the capillary wave model; this result was in contrast with assumptions made by other authors [43] to explain ion transport dynamics at this interface. Moreover, the calculations demonstrated the existence of 2-3 layers of water molecules whose dipoles were parallel to the interface. However, the broad orientational distribution of these layers made it quite unlikely that they could constitute a significant interfacial barrier to transport in that region. It was suggested that the rate-limiting step in the ion transport is the necessary switching of the solvation shell from one liquid to the other; the presence of a very rough interfacial structure may assist with this process. Finally, the DCE *gauche-trans* isomerization reaction was investigated at the interface and a continuum electrostatic model for the torsional potential of mean force was developed: this model was found to be only qualitatively adequate and solvent structure was determined to play a significant role although the sharp interface might have increased the merit of a continuum dielectric model.

2.3   Experimental studies

Experiments on electrolytes at liquid/air and liquid/liquid interfaces can be classified either as using indirect methods, such as interfacial tension measurements, or more direct ones, such as spectroscopic methods.

2.3.1 Interfacial tension measurements

Information on the properties of ions at liquid/air or liquid/liquid interfaces can be obtained experimentally through the determination of interfacial tensions. This conclusion is drawn from the Gibbs equation that relates excess amounts to interfacial tension (see the Section Capillarity).

Let us notice that, generally, the term "surface tension" refers to the liquid/air interface, while "interfacial tension" relates to liquid/liquid interfaces. However the latter should designate any type of interface.

There are at least two remarks that can be made in the first place. First, for a long time, these measurements have been questionable because of experimental difficulties. Clearly, measurements at low concentrations can be perturbed by small amounts of surface active impurities. Second, it might be said that, for a long time, experimentalists and theoreticians have had diverging viewpoints as to the origin of surface tension variation with concentration: generally, experimentalists have not considered the effect of image forces as the main responsible effect. So, the following statement [44] may illustrate this widespread opinion: "One must remember that, notwithstanding the immense sophistication of the mathematical treatments of electrolyte solutions by Onsager and coworkers, one here deals with another case where the highly developed mathematical treatment is applied only to an obviously oversimplified conceptual model. This model neglects, among other things, the detailed structural features of the solvent (the water), and this neglect can only result in a treatment which can hardly be any better than an approximation". Thus, in most experimental papers, the discussions about the variation of interfacial tension with concentration were expressed in terms of solvent effects rather than dielectric polarization effects of the OS type. In this respect, it is interesting to notice that the recent MD calculations described in the previous section have been developed in the same concern.

The experimental techniques used have been of various types, among which one may cite:

- The capillary rise method, generally considered to be one of the best and most accurate absolute methods, good to a few hundredths of a percent in precision. On the other hand, a zero contact angle on the wall of the capillary is required, which can be a practical limitation and a source of uncertainty.

- The maximum bubble pressure method, in which bubbles of an inert gas are blown in the liquid by means of a tube. This method is accurate to a few tenths of a percent and does not depend on contact angle.

- Detachment methods:

  1. The drop weight method is fairly accurate and very convenient for laboratory use. Its principle is to form drops of the liquid at the end of a tube until they fall; their weight is related to the surface tension. However, empirical corrections must be applied which limit the accuracy of this technique.

  2. The ring method, attributed to du Noüy, has been widely used. It involves the determination of the force necessary to detach a ring or loop of wire from the surface of the liquid, or interface between the liquids. This method is of good precision. A zero or near zero contact angle is necessary. Otherwise, a Teflon or polyethylene ring may be used.

- The Wilhelmy slide method requires no correction, in contrast with the previous techniques, and is simple to use. It is based on the principle that a thin plate immersed vertically in the liquid supports a meniscus whose weight, measured statically or by detachment, is given very accurately by the equation

$$W = W_{plate} + \gamma p \qquad (2.23)$$

  where p is the perimeter of the plate. This equation holds to within 0.1%. Another procedure is to level gradually the liquid until it just touches the plate. The increase in weight is then

$$W - W_{plate} = \gamma \, p \, \cos\theta \qquad (2.24)$$

  where θ is the contact angle that can be determined accurately.

- Methods based on the shape of static drops or bubbles include the pendant drop method and the sessile drop or bubble method. The general procedure is to make certain measurements of the dimensions or profile. It is accurate to a few tenths of a percent.

- Dynamic methods study the relaxation properties of interfaces. They include the flow methods and the study of capillary waves.

We now examine some of the experimental results for interfacial tension measurements.

"Anomalies"

We first evoke some peculiarities of surface tension that were, at a time, the subject of debate.

Jones-Ray effect.

These authors studied [45] the surface tension of very dilute solutions of electrolytes using the capillary rise method. They observed a lowering of surface tension of water upon adding KCl, K_2SO_4 and $CsNO_3$, with a minimum between 0.001 and 0.002 mol L^{-1}. The relative decrease amounted to about 2×10^{-4} for the three solutions. After having passed through the minimum, the surface tension increased as the concentration increased and attained the relative value of 1 at about 5.5×10^{-3} molar. For concentrations up to 0.01 molar the increase followed the OS theory.

This effect has been the subject of considerable controversy. Subsequent studies using various methods were not conclusive as to the reality of the effect. Langmuir has proposed [46] that the anomalous results were due to a change in the effective diameter of the silica capillary owing to a layer of electrolyte immediately adjacent to the capillary wall, related to the zeta potential. Onsager [47] made an extended theoretical investigation of this effect and developed a quantitative theory of the influence of the zeta-potential on the capillary rise. His theory accounted for most of the Jones and Ray results and led to an accurate determination of the zeta-potential of quartz. An attempt to interpret the results as a pure surface tension effect has also been made [48]. Besides, it has been suggested recently [9] that an initial decrease in the surface tension could be attributed to a nonzero solubility of the salt in the vapor.

Anomalies in the temperature dependence of the surface tension of water.

The dependence of the surface tension of pure water was found to exhibit anomalies [44] in the sense that the corresponding curve showed inflection points at several temperatures (15, 30, 45, 60°C,...). Anomalies, also called "kinks", were also found for other properties of pure water, such as viscosity, compressibility or magnetic susceptibility; besides, "polywater" has been the subject of much interest [49]. However, more careful studies have finally [38] pointed to the absence of "kinks" in the behavior of pure water; the anomalies described previously are now believed to have resulted from some artefacts or to have been within experimental error.

Concentration and temperature dependence

For the simple inorganic electrolytes early studies [7, 8] showed that in very dilute solutions equivalent concentrations of electrolytes of the same valence type gave similar surface tension increments $\Delta\gamma$, while at higher concentrations specific differences appeared. At concentrations below 0.01 molar $\Delta\gamma$ is very well described by the OS theory but, above it, the theoretical curve falls well below the experimental curves. At higher concentrations Schmutzer's equation 2.21 was found to account well for the experimental data up to 0.1 molar, with a layer thickness δ of 3.2 Å.

More recent studies have been concerned with measurements in concentrated solutions, of the order of 1-4 molar, with special care in the experimental measurements. A variety of electrolytes has been studied.

The surface tension of a range of alkali and alkali metal chlorides, ammonium chloride and tetramethyl ammonium chloride has been measured [50] at 21°C between 0 and 4 molar. The Wilhelmy technique with a mica or platinum plate was used. The agreement with available data, generally taken from the International Critical Tables (edition of 1928), was quite good. New values were obtained for CsCl, $(CH_3)_4NCl$, $MgCl_2$ and $CaCl_2$. The results were discussed in terms of the structure alteration caused by cations in the interfacial layer: a structure-breaker ion like Cs^+ was thought to be more easily accomodated in the interface than Li^+, a structure maker, due to the fact that it introduces more disorder in the structured region at the air-water interface.

In a similar work [18], the surface tension, and its temperature derivative $d\gamma/dT$, were measured by means of Wilhelmy slide techniques for dilute (< 1.5 molar) aqueous solutions of NaCl, KCl, NaI and KI. The data were used to determine the characteristics of a surface phase containing only the water component. The results showed that the distance of approach to the Gibbs dividing plane was approximately independent of the salt concentration and that it was smaller for the iodide ions than for the chloride ions due to differing modes of interaction with the surface zone water molecules: it was noticed that the iodide ion is larger than the chloride ion and it has a greater polarizability and lower polarizing ability; thus, the iodide ion can approach the Gibbs plane more closely because it can couple more readily with surface zone water molecules so as to compensate for its sterical disturbance of the surface-induced water structure.

Later, [51] the surface tensions of aqueous electrolytes and their interfacial tensions against n-dodecane were determined at 20°C using the drop weight method. The salts studied were LiCl, NaCl, KCl, KBr, NaBr, KI and Na_2SO_4. For the alkali metal chlorides and Na_2SO_4 the surface and interfacial tension increments ($d\gamma/dm$) were similar for a given electrolyte. The corresponding increments for KBr, NaBr and KI however were found to differ considerably. The results were discussed in terms of both electrostatic theory and dispersion force theory of interfaces. A salt-free layer of water was assumed near the interface and its thickness was estimated as previously [18]; both results were in good agreement when available. Apart from KI which is positively adsorbed at the dodecane/water interface, all the salts were negatively adsorbed at the liquid/vapor and the liquid/liquid interfaces. The difference between the interfacial tension increment and the surface tension increment for a given electrolyte was equated with the work of adhesion between the water and the alkane phases; this work was estimated in the framework of the dispersion forces theory of Israelachvili [52, 53] for the van der Waals attraction between two macroscopic media.

Further surface tension measurements [54] have been conducted for aqueous solutions of some 1:1 electrolytes: NaF, KF, NaNO$_3$ and KNO$_3$, between 280 and 310 K up to 1.5 molar, by means of the ring detachment method. The data were interpreted by considering the free energy of hydration of the salt in the interfacial region: there, the lower number-density of water molecules leads to a free energy of hydration that is less negative than in the bulk, yielding a depletion of solute in the surface region. The extent of this depletion was expected to correlate with the bulk thermodynamic parameters of hydration. A simplified but reasonable calculation led the authors to plot the quantity $d\gamma/da_\pm$, with $a_\pm$ the mean activity of the salt in the bulk, against the standard enthalpy of hydration of the electrolyte: the scatter of points was rather large but a significant tendency to correlate emerged from the plot.

Quite recently, systematic surface tension measurements have been performed [55] on 34 electrolytes of 1:1, 1:2, 1:3 and 2:2 valence type, between 0 and 1 molar at room temperature, using the maximum bubble pressure method. The experiments were concerned with the determination of the surface tension gradient $d\Delta\gamma/dC$ (with C the electrolyte concentration), which was found to vary between 20 and 30°C within the experimental error. For this reason the temperature was not precisely controlled. Graphs of surface tension relative to water vs. electrolyte concentration gave straight lines, of slope $d\Delta\gamma/dC$. The results gave positive and negative values of $d\Delta\gamma/dC$. Negative values were found for HCl, HNO$_3$, HClO$_4$, CH$_3$COOH, CH$_3$COO(CH$_3$)$_4$N and mixtures of NaCl and HClO$_4$. The electrolytes were classified according to their $d\Delta\gamma/dC$ value: if the absolute value of $d\Delta\gamma/dC$ was greater than 1 then both ions of the electrolyte were classified as either positively or negatively adsorbed; electrolytes for which the cation was negatively adsorbed and the anion was positively adsorbed, or *vice versa*, were considered to give a $d\Delta\gamma/dC$ value between -1 and +1. The value of ± 1 was chosen arbitrarily to distinguish between a strong (< -1 or $> +1$) or a weak (-1 to +1) affinity of the ions as an ion-pair for the interface. No such systematic study had been reported before. The trends in $d\Delta\gamma/dC$ were examined for a wide range of chloride electrolytes and the standard molar entropies of hydration of the countercations were plotted against $d\Delta\gamma/dC$: a very good correlation was observed. Entropy of hydration was regarded as being an indicative and sensitive measure of the extent of ionic hydration. So, the negative adsorption of strongly hydrated cations (and anions) from the interface was explained by the hydration free energy dominating the free energy required for a "bare" ion to exist in the bulk solution or at the interface. Positive adsorption was rather explained by forces such as van der Waals forces or hydrophobic attraction;

for example, the gas/water interface is negatively charged (with a potential of about -15 mV) and behaves like a hydrophobic surface [56]; hence cations and ions with hydrophobic functionality (e.g. acetate and tetramethylammonium ions) may be attracted toward the gas/water interface.

Subsequently the same authors have reported [57] on their experiments with more details, in the same experimental conditions. A correlation was found between $d\Delta\gamma/dC$ and the Jones-Dole viscosity coefficient B defined by the relation

$$\eta/\eta_W = 1 + A\sqrt{C} + BC + ...$$

in which the coefficient A is due to interionic electrostatic forces and the coefficient B is representative of the retardation of solution flow due to hydration of ions. Besides a good correlation was found between $d\Delta\gamma/dC$ and the exponential decay coefficient for oxygen solubility in 8 cationic chloride solutions (oxygen solubility decreases exponentially with increasing electrolyte concentration).

Direct probes of interfaces

Development of several experimental techniques has recently allowed to probe liquid/air and liquid/liquid interfaces on a molecular scale.

Among these techniques, particularly powerful are the sum frequency generation SFG [63, 64, 65] and second harmonic generation (SHG) [66, 67, 68] techniques:

- SFG is a nonlinear optical method which was pioneered by Shen [65]. It is based on the second-order optical phenomenon (see Figure 2.7) which consists of illuminating the surface of a solid or liquid with two overlapping pulsed laser beams of different frequencies ω_1 and ω_2. At the surface, the intense optical fields induce a second-order polarization of the medium at $\omega_1 \pm \omega_2$. The latter results in the production of a coherent optical field, which for SFG is at the frequency $\omega_1 + \omega_2$. The SF light is produced in both reflectance and transmittance at angles a few degrees from the linear reflection and transmission angles.

- Similarly, SHG is the non linear conversion of two photons of the same frequency ω to a single photon of frequency 2ω.

A related method, which seeks to minimize the excitation of bulk molecules, involves an evanescent wave technique, *i.e.* total internal reflection (TIR). Generally,

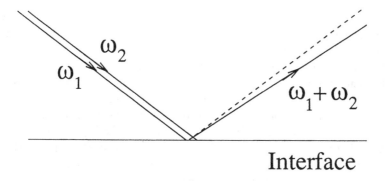

Figure 2.7: Principle of sum-frequency generation experiments.

this method is not interface specific because the evanescent wave penetrates to a distance of the order of half the light wavelength into the bulk solution. However, when the adsorption of a probe molecule to the interface is so strongly favored that there is only a very small bulk concentration the signal originating from the bulk molecules can be neglected. This property has been used in particular for a TIR-SFG method; the maximum intensity then occurs at the critical angles for both incoming fundamental beams.

The reason that SFG and SHG can selectively probe interfaces, without being overwhelmed by bulk species, is that these second-order processes are electric dipole forbidden in centrosymmetric media. So, the ability of noncentrosymmetric crystals to produce SHG has led to their implementation as frequency doublers in a variety of laser systems. SFG and SHG can be produced by the break in symmetry that occurs at the interface between two centrosymmetric media. Only the first few atomic or molecular layers on either side of the interface participate in this symmetry breaking because they experience different interactions in the "up" vs the "down" directions. Thus, these techniques can be used as a highly surface-selective optical probe of interfaces.

Liquid/air interface

Neutron reflectivity with isotopic substitution has been used [58] to determine the structure of triethylene glycol monododecyl ether adsorbed at the air/water interface. Besides, X-ray diffraction and reflection studies [59] have provided structural details on an angstrom level about molecular packing at the interface. Optical techniques such as Fourier transform-infrared [60], Brewster angle microscopy [61], fluorescence microscopy [62], SFG [63, 64, 65] and SHG [66, 67, 68] have provided information about the phase behavior and structure and orientation of molecules.

An interesting application of optical techniques has been the study of chemical reactions at the air/water interface. So, an acid-base reaction has been investigated: using SHG, *p*-nitrophenol (denoted by HA) was shown [69] to be much less dissociated in the interfacial region; it was estimated that the ratio of A^- to HA was at least 50-100 times smaller than in the bulk solution. The same tendency was found in subsequent studies for hexadecylanilinium [70], showing that the interface favors the neutral over the charged form in acid-base equilibria. Besides, the dynamics of a photoisomerization reaction has been found [71] to be faster at the interface than in the bulk liquid, which phenomenon was suggested to be associated with lower friction at the interface for motion along the isomerization coordinates.

The adsorption kinetics of a long-chain surfactant has been followed [72] by measuring the time dependence of a SHG signal. The kinetics did not obey a $t^{1/2}$ dependence, but the data could be fit to a Langmuir adsorption model.

Liquid/liquid interface

Progress has been much slower in understanding the molecular structure at the interface between two bulk liquids because of inapplicability or lack of surface specificity of most surface techniques for probing this interface.

Neat alkane/water interfaces have been investigated. For example, a series of neat *n*-alkane/water interfaces has been studied by TIR-SHG [73]. Surface nonlinear susceptibility measurements suggested a significant degree of molecular ordering. Surprisingly, a higher degree of ordering was found for the alkane/water systems consisting of even numbered hydrocarbon chains relative to the odd numbered hydrocarbons.

Studies have been done to investigate the orientation and adsorption of surface active dyes at the oil/water interface using fluorescence [74, 75], resonance Raman scattering [76, 77]. Neutron reflectivity has been used [78] to determine the thickness of a surfactant (monodecyl tetraglycol ether) layer at the octane/water interface and the interfacial roughness, which was large (*ca.* 90 Å) due to the very low interfacial tension ($\gamma = 0.08$ mN m^{-1}).

The molecular orientation of surfactants has been determined by SHG: the first detailed study was performed [79] for the adsorption of sodium 1-dodecylnaphtalene-4-sulfonate (SDNS) at the aqueous/decane and aqueous/CCl$_4$ interfaces; the results showed that the molecular orientation depends on the nature of the nonaqueous phase. Recent advances have used SHG to probe surfactants containing an aromatic head group at the water/dichloroethane interface in an electrochemical cell [80];

by varying the potential across the cell the population of the charged form of the surfactant at the interface could be varied; measurement of the magnitude of the SH signal showed that the interface population did vary with potential, while measurement of the polarization of the SH light showed that the surfactant orientation did not change with the applied potential.

A similar study has been conducted using TIR-SFG [81] on the conformational order of sodium dodecyl sulfate (SDS) adsorbed at the D_2O/CCl_4 interface. A change in conformation of the alkyl chain with increased surface coverage was observed. Polarization studies indicated that the terminal methyl group axis was oriented primarily along the surface normal.

SFG studies have also allowed the description of the structure of interfacial water and how it is affected by the presence of charged surfactants [64, 82]. The vibrational structure of both the interfacial water molecules and adsorbed surfactant were probed in these studies as increasing amounts of surfactant were added to the aqueous phase. The most striking result occurred in the O-H stretching region corresponding to interfacial water: with increased surfactant concentration, a strong enhancement in the ice-like peak was observed relative to the surfactant-free interface. This enhancement has been attributed [83] to increased orientation of water molecules in the double-layer region, which is induced by the large electrostatic field created by the charged surfactant and counterion. Ionic strength studies have supported this conclusion: increasing amounts of NaCl have been shown to decrease the response from the O-H band, which could be ascribed to a screening effect that limited the number of interfacial water molecules affected by the electrostatic field [82]. The water molecules were found to align in opposite directions depending on the cationic or anionic character of the surfactant.

Bibliography

[1] J. Lyklema, *Fundamentals of Interface and Colloid Science*, Academic Press, London, 1991.

[2] A.W. Adamson, *Physical Chemistry of Surfaces*, Wiley-Interscience, New York, 1990.

[3] C. Wagner *Physik. Z.* 25 (1924) 474.

[4] L. Onsager and N.N.T. Samaras *J. Chem. Phys.* 2 (1934) 528.

[5] F.P. Buff and F.H. Stillinger *J. Chem. Phys.* 25 (1956) 312.

[6] M.A. Wilson, A. Pohorille and L.R. Pratt *Chem. Phys.* 129 (1989) 209.

[7] G. Schwenker *Ann. Physik* 11 (1931) 525.

[8] A. Heydweiller *Ann. Physik* 33 (1910) 145.

[9] A.L. Nichols and L.R. Pratt *J. Chem. Phys.* 80 (1984) 6225.

[10] R.A. Stairs *Can. J. Chem.* 73 (1995) 781.

[11] E. Schmutzer *Z. Phys. Chem. (Leipzig)* 204 (1955) 131.

[12] J.G. Kirkwood *J. Chem. Phys.* 3 (1935) 300.

[13] J.G. Kirkwood *Chem. Revs.* 19 (1936) 275.

[14] F.P. Buff *Z. Elektrochem.* 56 (1952) 311.

[15] F.P. Buff *J. Chem. Phys.* 23 (1955) 419.

[16] T. Croxton, A. Mc Quarrie, G.N. Patey, G.M. Torrie and J.P. Valleau *Can. J. Chem.* 59 (1981) 1998.

[17] T. Croxton and A. McQuarrie *J. Phys. Chem.* 87 (1983) 3407.

[18] K. Johansson and J.C. Eriksson *J. Colloid Interface Sci.* 49 (1974) 469.

[19] A.L. Nichols and L.R. Pratt *J. Chem. Phys.* 76 (1982) 3782.

[20] M.A. Wilson, A.L. Nichols and L.R. Pratt *J. Chem. Phys.* 78 (1983) 5129.

[21] R.M. Mazo and C.Y. Mou in *Activity Coefficients in Electrolyte Solutions*, R.M. Pytkowicz Ed., CRC Press, Boca Raton, Florida, 1983.

[22] S. Levine and C.W. Outhwaite *J. Chem. Soc. Faraday Trans. II* 74 (1978) 1670.

[23] S. Levine, C.W. Outhwaite and L.B. Bhuiyan *J. Electroanal. Chem.* 123 (1981) 105.

[24] J.G. Kirkwood *J. Chem. Phys.* 2 (1934) 767.

[25] L.B. Bhuiyan, D. Bratko and C.W. Outhwaite *J. Phys. Chem.* 95 (1991) 336.

[26] L.B. Bhuiyan, D. Bratko and C.W. Outhwaite *J. Phys. Chem.* 90 (1986) 6248.

[27] N.I. Christou, J.S. Whitehouse, D. Nicholson and N.G. Parsonage *Mol. Phys.* 55 (1985) 397.

[28] M.A. Wilson, A. Pohorille and L.R. Pratt *J. Phys. Chem.* 91 (1987) 4873.

[29] M.A. Wilson, A. Pohorille and L.R. Pratt *J. Chem. Phys.* 88 (1988) 3281.

[30] M.A. Wilson, A. Pohorille and L.R. Pratt *J. Chem. Phys.* 90 (1989) 5211.

[31] M. Matsumoto and Y. Kataoka *J. Chem. Phys.* 88 (1988) 3233.

[32] M.C. Goh and K.B. Eisenthal *Chem. Phys. Lett.* 157 (1989) 101.

[33] M.A. Wilson, A. Pohorille and L.R. Pratt *J. Chem. Phys.* 88 (1988) 3281.

[34] L.R. Pratt *J. Phys. Chem.* 96 (1992) 25.

[35] W.L. Jorgensen, J. Chandresekhar, J.D. Madura, R.W. Impey and M.L. Klein *J. Chem. Phys.* 83 (1985) 5832.

[36] J. Chandresekhar, D.C. Spellmeyer and W.L. Jorgensen *J. Am. Chem. Soc.* 106 (1987) 903.

[37] M.A. Wilson and A. Pohorille *J. Chem. Phys.* 95 (1991) 6005.

[38] J.E.B. Randles *Phys. Chem. Liq.* 7 (1977) 107.

[39] M.E. O'Neill and K.B. Ranger *Phys. Fluids* 26 (1983) 2035.

[40] I. Benjamin *J. Chem Phys.* 95 (1991) 3698.

[41] M. Meyer, M. Mareschal and M. Hayoun *J. Chem. Phys.* 89 (1988) 1067.

[42] Benjamin, I. *J. Chem. Phys.* 97 (1992) 1432.

[43] Y. Shao and H.H. Girault *J. Electroanal. Chem.* 282 (1990) 59.

[44] W. Drost-Hansen *Ind. Eng. Chem.* 57 (1965) 18.

[45] G. Jones and W.A. Ray *J. Am. Chem. Soc.* 59 (1937) 187; *ibid.* 63 (1941) 288.

[46] I. Langmuir *Science* 88 (1937) 430; *ibid. J. Chem. Phys.* 6 (1938) 873.

[47] L. Onsager and E. Drauglis in *Quantum Statistical Mechanics in the Natural Sciences*, Eds. S.L. Mintz and al., Plenum Press, New York, 1974.

[48] M. Dole *J. Am. Chem. Soc.* 60 (1938) 904.

[49] Symposium on Polywater *J. Colloid Interface Sci.* 36 (1971) 415.

[50] J. Ralston and T.W. Healy *J. Colloid Interface Sci.* 42 (1973) 629.

[51] R. Aveyard and S.M. Saleem *J. Chem. Soc. Faraday Trans.* 72 (1976) 1609.

[52] J.N. Israelachvili *Proc. Roy. Soc. A* 331 (1972) 39.

[53] J.N. Israelachvili *J. Chem. Soc. Faraday II* 69 (1973) 1729.

[54] M.J. Hey, D.W. Shield, J.M. Speight and M.C. Will *J. Chem. Soc. Faraday Trans. I* 77 (1981) 123.

[55] P.K. Weissenborn and R.J. Pugh *Langmuir* 11 (1995) 1422.

[56] W.A. Ducker, Z. Xu, and J.N. Israelachvili *Langmuir* 10 (1994) 3279.

[57] P.K. Weissenborn and R.J. Pugh *J. Colloid Interface Sci.* 184 (1996) 550.

[58] J.R. Lu, E.M. Lee, R.K. Thomas, J. Penfold and S.L. Flitsch *Langmuir* 9 (1993) 1352.

[59] M.C. Shih, T.M. Bohanon, J.M. Mikrut, P. Zshack and P. Dutta *Phys. Rev. A* 45 (1992) 5374.

[60] J.T. Buontempo and S.A. Rice *J. Chem. Phys.* 98 (1993) 5835.

[61] B. Fischer, M.W. Tsao, J. Ruiz-Garcia, T.M. Fischer, D.K. Schwarz and C.M. Knobler *J. Phys. Chem.* 98 (1994) 7430.

[62] X. Qiu, J. Ruiz-Garcia, K.J. Stine, C. Knobler and J.V. Selinger *Phys. Rev. Lett.* 67 (1991) 703.

[63] G.R. Bell, C.D. Bain and R.N. Ward *J. Chem. Soc. Faraday Trans.* 92 (1996) 515.

[64] D.E. Gragson, B.M. McCarty and G.L. Richmond *J. Phys. Chem.* 100 (1996) 14272.

[65] Y.R. Shen *Nature* 337 (1989) 519.

[66] G.L. Richmond, J.M. Robinson and V.L. Shannon *Prog. Surf. Sci.* 28 (1988) 1.

[67] R.M. Corn and D.A. Higgins *Chem. Rev.* 94 (1994) 107.

[68] K.B. Eisenthal *Chem. Rev.* 96 (1996) 1343.

[69] K. Bhattacharyya, E.V. Sitzmann and K.B. Eisenthal *J. Chem. Phys.* 87 (1987) 1442.

[70] X. Zhao, S. Subrahmanyan and K.B. Eisenthal *Chem. Phys. Lett.* 171 (1990) 558.

[71] E.V. Sitzmann and K.B. Eisenthal *J. Phys. Chem.* 92 (1988) 4579.

[72] T. Rasing, T. Stehlin, Y.R. Shen, M.W. Kim and P. Valint *J. Chem. Phys.* 89 (1988) 3386.

[73] J.C. Conboy, J.L. Daschbach and G.L. Richmond *J. Phys. Chem.* 98 (1994) 9688.

[74] M.J. Wirth and J.D. Burbage *J. Phys. Chem.* 96 (1992) 9022.

[75] D.A. Piasecki and M.J. Wirth *J. Phys. Chem.* 97 (1993) 7700.

[76] T.T. Takenaka and T. Nakanaga *J. Phys. Chem.* 80 (1976) 475.

[77] Y. Tian, J. Umemura and T.T. Takenaka *Langmuir* 4 (1988) 1064.

[78] L.T. Lee, D. Langevin and B. Farnoux *Phys. Rev. Lett.* 67 (1991) 2678.

[79] S.G. Grubb, M.W. Kim, T. Rasing and Y.R. Shen *Langmuir* 4 (1988) 452.

[80] D.A. Higgins and R.M. Corn *J. Phys. Chem.* 97 (1993) 489.

[81] J.C. Conboy, M.C. Messmer and G.L. Richmond *J. Phys. Chem.* 100 (1996) 7617.

[82] D.E. Gragson, B.M. McCarthy and G.L. Richmond *J. Am. Chem. Soc.* 119 (1997) 6144.

[83] G.L. Richmond *Anal. Chem.* 69 (1997) 536A.

Chapter 3

Solute transfer kinetics at a liquid/liquid interface

Solute transfer kinetics across the interface between two immiscible liquids has relevance in several areas of physical chemistry, chemical engineering and biology. Examples include ion extraction at free [1] or polarized interfaces [2, 3] and ion transport in biological membranes [4, 5, 6].

This chapter is concerned with the study of kinetics of solute transfer at free liquid/liquid boundaries. This process has found many applications in separation science [1] with developments in hydrometallurgy (e.g. Cobalt/Nickel separation), nuclear fuel reprocessing, pharmaceutical industry and supported liquid membrane technology. indexTransfer kinetics

The literature is now abundant in this field. However, the mechanisms of the extraction processes are still the subject of speculation and controversy. Nevertheless, this situation is probably beginning to change as new experimental techniques are now able to study liquid interfaces on a molecular scale: optical second harmonic and sum frequency generation (SHG and SFG) are described in chapter 2 of this book. In particular, these techniques may have the potential to obtain dynamical information in the future, as has been demonstrated already for the water/air interface [7].

The aim of this chapter is to give an overview of this subject, with focus on the fundamentals of liquid/liquid extraction kinetics. This chapter will not examine the chemical and practical aspects of the problem. The reader is referred to other publications (e.g. [1, 8]) for detailed accounts of these points.

3.1 Basic aspects of the problem

Extraction systems

There are roughly two classes of extraction systems. The situtation is depicted in Figure 3.1.

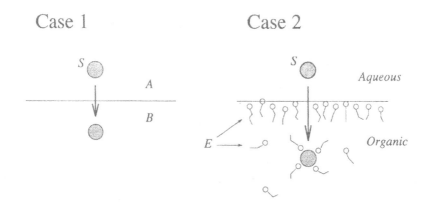

Figure 3.1: Simplified picture showing the two categories of liquid/liquid extraction systems. Case 1: Homogeneous phases A and B; Case 2: An extractant E is present in the organic phase to solubilize species S in organic B.

In the first case, transfer consists simply of a desolvation-resolvation process of the solute at the interface between two homogeneous phases. In the second case, the solute is solubilized in the organic phase by an extractant E added to a diluent; generally the extractant has amphiphilic, surface active, properties that enhance its extracting properties, and the solute is not soluble in the diluent. In both cases the species is generally in an electrically neutral form in the organic phase, in most cases where the latter has a low dielectric permittivity. Therefore, if the species is an ion, it will either be complexed by the extractant in its ionized form or associate with a counterion taken from the aqueous phase.

To the first class belong systems such as alkanoic acids, methylnicotinate, salicylic acid and urea extracted by liquid hydrocarbons or big esters such as isopropyl-tetradecanate (isopropyl-myristate (IPM)).

The second class concerns a wide variety of compounds and extractants, among which one may quote the following systems: metal cations such as Cobalt(II), Nickel(II), Zinc(II) or lanthanides extracted by di-(2-ethylhexyl) phosphoric acid denoted by HDEHP or D2EHPA, a widely used extractant for industrial applications; and nitric acid HNO_3 or uranyl ion extracted by tributylphosphate (TBP).

Interfacial reaction vs. bulk transport

The transfer process of a solute S from a liquid A to a liquid B consists at least of 3 steps:

1. Transport of S from the bulk of A to the interface region.

2. A-to-B interfacial transfer.

3. Transport of S from the interface to the bulk of B.

which are depicted on Figure 3.2.

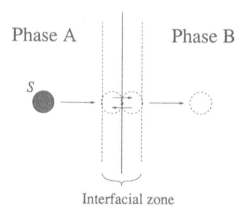

Interfacial zone

Figure 3.2: Overall transfer process, consisting of at least 3 steps (transport from the bulk of phase A to interface, reaction in interfacial zone and transport to the bulk of B).

Therefore, the interfacial kinetics (step 2) may be determined by subtracting the transport contributions 1 and 3 from the overall process, or by making them very small.

Two limiting cases may be encountered in practical situations:

- The transport contribution is very small in comparison with the chemical reactions contribution. Then the system is said to be in a kinetic regime, or kinetically controlled, and the process is limited by the kinetics of the chemical reaction(s).

- The opposite case corresponds to a diffusional regime.

3.2 Basic modelling of extraction process

Rate constant and characteristic reaction time

Consider a very thin lamella of a liquid A formed on a flat solid (Figure 3.3). This phase contains a species S which is allowed to be transferred to another phase B. At the interface A/B the extraction process is supposed to be irreversible, and the lamella is so thin that the process is assumed to be controlled by the interfacial transfer.

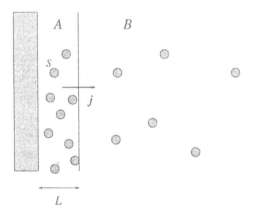

Figure 3.3: Transfer of species S at the interface between A and B, through an irreversible reaction which controls the process.

The concentration of the species is supposed to be sufficiently low, so that the interfacial flux density (the number of moles of S crossing a surface of unit area) may be written

$$j = k\,C \tag{3.1}$$

where C is the molar concentration of S in A and k is the interfacial reaction rate constant.

This equation shows that k has the dimension of m s^{-1}. Moreover, k must be a constant as long as C is not too high. Above some critical concentration eq 3.1 can be expected not too hold any longer because of nonlinearities due to interactions between solute particles S at, or near, the interface.

Thus the equation verified by C for such a reaction is

$$V\frac{dC}{dt} = -j\ S_{int} \qquad (3.2)$$

with V the volume of A and S_{int} the interfacial area. Since $V = LS_{int}$, where L is the thickness of the lamella, we get from eqs 3.1 and 3.2

$$\frac{dC}{dt} = -\frac{k}{L}\ C \qquad (3.3)$$

from which the characteristic reaction time can be defined as

$$\tau = L/k \qquad (3.4)$$

which shows that, for such an heterogeneous reaction, the characteristic time is related to the size of the donor phase.

Sequential process in stationary conditions

As mentioned above, the global extraction process consists of a series of steps involving transport and chemical reactions. This type of process is sequential by nature.

Some systems operate in a stationary regime. In the more simple case, the solute must cross a diffusive barrier in the donor phase A, then the interface and lastly another diffusive barrier in B. This is for instance the case of the Lewis cell described below: both phases are stirred but more "quiescent" liquid layers exist near the interface. The process is described in Figure 3.4.

Very often, transport in a liquid near the interface is modelled as a purely diffusive process in a stagnant layer of some thickness δ. Generally this thickness has different values δ_A and δ_B in phases A and B. The concept of stagnant layer was introduced by Nernst at the beginning of this century and it constituted the basis of the two-film theory proposed by Whitman[9]. The parameter δ represents the

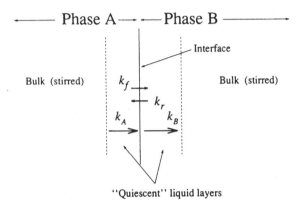

Figure 3.4: Transfer of solute from A to B, through liquid layers in A and B, plus interfacial reaction.

diffusion layer thickness for the particular type of diffusive transport process created near the interface. However, the main difficulty associated with this quantity is that it cannot be estimated independently as a function of the physico-chemical parameters of the system.

Application of first Fick's law gives the corresponding flux density

$$j = D \frac{\Delta C}{\delta} \tag{3.5}$$

with D the diffusion coefficient and because, in stationary state, the concentration profile is linear in this unstirred liquid layer and the concentration gradient $\partial C/\partial x$ can be replaced by $\Delta C/\delta$ with ΔC the concentration drop on the distance δ.

Comparison of eqs 3.1 and 3.5 shows that diffusive transport can mimic [10, 11] an interfacial reaction, of equivalent rate constant

$$k_{diff} \equiv \frac{D}{\delta} \tag{3.6}$$

In the following we will denote by k_A and k_B the two corresponding "diffusive" rate constants in A and B, that is

$$k_A \equiv \frac{D_A}{\delta_A} \quad \text{and} \quad k_B \equiv \frac{D_B}{\delta_B} \tag{3.7}$$

where D_A, D_B and δ_A, δ_B stand for the diffusion coefficients and diffusion layer thicknesses in A and B, respectively.

Then, for the stationary process described in Figure 4, the flux density may be expressed as

$$j = k_A(C_A - C_{A,i}) = k_f C_{A,i} - k_r C_{B,i} = k_B(C_{B,i} - C_B) \tag{3.8}$$

with C_X the concentration of species S in the bulk of phase X= A or B; $C_{X,i}$ the concentration of S in phase X, close to the interface; k_f and k_r the forward (A→B) and reverse (B→A) rate constants for the reactions occurring at the interface.

The latter equation can also be written as

$$j = k_{A \to B} C_A - k_{B \to A} C_B \tag{3.9}$$

in which $k_{A \to B}$ and $k_{B \to A}$ denote the forward and reverse effective rate constants for the processes involving both interfacial reaction and transport, which are the observed rate constants in a practical experiment.

By making successively $C_B \equiv 0$ and $C_A \equiv 0$ eqs 3.8 and 3.9 can be easily solved to yield

$$k_{A \to B}^{-1} = k_A^{-1} + k_f^{-1} + \frac{1}{K} k_B^{-1} \tag{3.10}$$

and

$$k_{B \to A}^{-1} = k_B^{-1} + k_r^{-1} + K k_A^{-1} \tag{3.11}$$

with K the B-to-A partition coefficient defined through

$$K \equiv \frac{C_B^{eq}}{C_A^{eq}} \tag{3.12}$$

with the superscript "eq" denoting an equilibrium value.

By virtue of eqs 3.8 and because $j = 0$ at equilibrium it follows that

$$K = \frac{k_f}{k_r} \tag{3.13}$$

so that eqs 3.10 and 3.11 can be rewritten as

$$k_{A \to B}^{-1} = k_f^{-1} (1 + k_f/k_A + k_r/k_B) \tag{3.14}$$

and

$$k_{B \to A}^{-1} = k_r^{-1} (1 + k_f/k_A + k_r/k_B) \tag{3.15}$$

Eqs 3.10 and 3.11 are sometimes interpreted in terms of resistance to transfer. So, the global resistance to transfer from the bulk of A to the bulk of B is defined by

$$R_{A \to B} \equiv k_{A \to B}^{-1}$$

and it is the sum of the resistances of the successive steps of the process:

$$R_{A \to B} = R_A + R_f + R_B \tag{3.16}$$

where the resistance due to transport in B includes the distribution ratio K.

An important consequence of eqs 3.14, 3.15 is that $k_{A \to B}$ and $k_{B \to A}$ may be equated with k_f and k_r, respectively, if the following conditions are fulfilled

$$k_f \ll k_A \qquad \text{and} \qquad k_r \ll k_B \tag{3.17}$$

in which case the system can be said to operate in a regime controlled by the kinetics of the chemical reaction occurring in the interface region.

3.3   Experimental techniques

3.3.1   Two categories

The various available methods may be grouped into two categories:

- Those techniques which allow direct contact of the phases. Usually, the phases are stirred to accelerate mass transfer. Unfortunately, the influence of transport in either phase is poorly known. For this reason, it has been tried to minimize the transport contribution by vigorous stirring, to reach a regime controlled by the kinetics.

- The techniques which try to palliate this shortcoming by imposing controlled hydrodynamic conditions. This imposes the use of a third medium, generally a membrane containing one of the phases.

3.3.2   The techniques

Among the available techniques one may cite the following ones:

Category 1 includes the Lewis cell [12] and its modifications (e.g. the Armollex cell [13] and the Nitsch cell [14]), the highly stirred tanks (e.g. the AKUFVE apparatus [15]) and the moving drop technique [16].

To category 2 belong the rotating diffusion cell (RDC) [17], the short-time phase-contacting method [18], the rotating stabilized cell (RSC) [21] and the rotating membrane cell (RMC) [22, 23].

Those methods which belong to the first category suffer mainly from a lack of control of the hydrodynamics. Moreover, it is difficult [8, 24] to determine the interfacial area for highly stirred tanks in which small size droplets of one phase are produced inside the other phase. Generally, the first difficulty has been circumvented by the use of the following criterion: it is observed in many cases that the extraction rate increases as stirring is increased and then levels off above some stirring rate; the plateau value reached by the extraction rate is then interpreted as the limiting rate obtained in a kinetic regime. Here, it is intuitively assumed that the resistance to mass tranfer due to transport can be reduced to zero by increasing agitation. However stirring is in fact produced by propellers and identifying the degree of stirring with the rotation speed of these elements can be misleading: it has been underlined [25] that efficiency of stirring can be lost due to a "slip effect" of the fluids on the blades of the propellers. Besides, further possible side effects (such as droplet coalescence and insufficient internal drop circulation with the use of highly stirred tanks [8]) lead to the conclusion that "a plateau value in the extraction rate vs. stirring speed can be originated by physical phenomena which have nothing to do with the occurrence of a kinetic regime" [8]. Besides, Hughes and Rod [26] have shown formally that a plateau region may arise from mass transfer in which chemical kinetics and diffusion control the process. This empirical criterion therefore appears questionable. Other criteria have been proposed, which have been reviewed recently [27].

Difficulties associated with the methods of the second category may originate from the use of a membrane, which introduces a third constituent in the system. This membrane must interact as little as possible with the species that constitute the system: it must be chemically inert, exhibit low adsorption properties and it must not interfere with the extraction process at the interface.

Lewis-type cells

Figure 3.5 shows a sketch of a Lewis-type cell, which is probably still the most widely used technique.

In this type of cell the liquids are in direct contact. They are stirred independently, in turbulent motion and care is taken for not breaking up the interface. So, the interfacial area is well defined. Extraction is followed by sampling the phases. When this type of cell is used the data are sometimes described in the framework of the two-film theory [9, 27], which assumes that two stagnant layers of fluid exist on either side of the interface. However, the thickness of these layers cannot be calculated theoretically because the liquids are in turbulent motion, with undefined

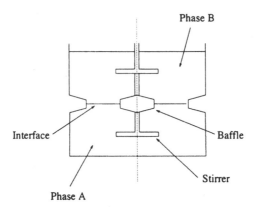

Figure 3.5: Cross section of the original Lewis cell.

hydrodynamic characteristics. Besides, it seems [24] that no empirical approach has been tried to correlate the concentration boundary layer thickness to measurable system parameters. Other phenomenological descriptions have been used such as the penetration theory [28, 29], which was developed in an attempt to account better for the turbulent nature of the flow.

Generally the extraction process has been described by rate laws of the type of eq 3.9 in which the interfacial reaction was assumed to be the rate-determining step, so leading to

$$V_A \frac{dC_A}{dt} = -Q\left(k_f C_A - k_r C_B\right) \tag{3.18}$$

with V_A the volume of phase A and Q the interfacial area. This relation shows that the rate of variation of C_A is proportional to the specific interfacial area

$$\bar{a} = Q/V_A \tag{3.19}$$

Rotating diffusion cell (RDC)

This method has been proposed by Albery in 1974 [30] and improved later [17]. A diagram of the RDC is given in Figure 3.6. The RDC has always been operated with laminar flow in the phases.

The flow is created by the rotation of a microporous membrane, impregnated with one phase and mounted on a hollow cylinder, at a definite speed. The purpose of the baffle is to prevent the inner solution from rotating with the motion

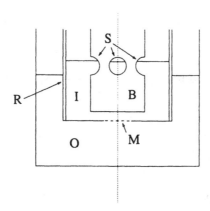

Figure 3.6: Cross section of the RDC: I= inner chamber; O= outer chamber; R= rotating perspex cylinder; B= baffle; S= slot; M= membrane.

of the cylinder. Mass transfer occurs from the inside of the cylinder, through the membrane, and into the outer liquid. The method offers the possibility of imposing various configurations, depending on which phase (aqueous or organic) is placed within the filter and in the two compartments.

The RDC has constituted the first decisive attempt to control the hydrodynamics. This control was achieved by adaptation of the rotating-disk electrode used in electrochemistry. Description of the hydrodynamic motion of a fluid near a rotating disk was first approached by von Karman in 1921 [31] and it was reconsidered then by Levich [32], who extended this theory to convective-diffusive transport to a rotating disk. The Levich theory has been analyzed in detail subsequently and corrections have been proposed [33, 34], which however are commonly very small in practice.

Expressions have been derived [17] for the global rate constant of the process in stationary conditions. The "sandwich" configuration corresponds to the case where the membrane contains the organic phase with the same aqueous solutions on either side of the membrane. For this configuration, the global resistance was found to be [17]

$$k^{-1} = 2k_A^{-1} + 2(\sigma k_f)^{-1} + (\sigma k_m)^{-1} \qquad (3.20)$$

in which σ designates the membrane surface porosity (the ratio of the area of the pores to the membrane area), A is the aqueous phase and B is organic, k_A^{-1} is the resistance of the diffusion layer and k_A is given by eq 3.6

$$k_A = D_A/\delta_A \qquad (3.21)$$

with D_A the diffusion coefficient of the species in A and the diffusion layer[1] thickness δ_A expressed by [32]

$$\delta_A = 1.612 \, (\nu_A/\omega)^{1/2} \, Sc_A^{-1/3} \tag{3.22}$$

with ν_A the kinematic viscosity of A, ω the rotation speed (in rad s^{-1}) and Sc$_A$ the Schmidt number

$$Sc_A = \nu_A/D_A \tag{3.23}$$

Moreover, in eq 3.20, k_m^{-1} is the membrane resistance defined by

$$k_m = L_{eff}/D_B \tag{3.24}$$

with L_{eff} the effective path length [35, 23], taking into account the effect of membrane tortuosity.

It has been pointed out [24, 27] that although "the inner chamber is baffled to promote rotating-disk hydrodynamics, no studies have shown conclusively that the flow inside the cylinder is the same as flow outside the cylinder", in which rotating-disk hydrodynamics is created. Besides, the main difficulties with this type of technique are associated with the use of the filter. They can be of physical or chemical nature: so, in the original work of Albery [17] and other subsequent studies [35, 36], L has been approximated by the membrane thickness; this parameter can be determined by a conductimetric method [23]. The porosity, for which the manufacturer generally gives an approximate value, can be measured independently. Besides, in most studies using the RDC [17, 35, 37, 38, 39], a Millipore MF membrane composed of mixed cellulose esters (nitrate and acetate) was used. This type of membrane is known to be poorly compatible with some chemical species, such as esters, organic acids and strong concentrated acids; precisely the ester IPM and the organophosphoric acid HDEHP have been used in experiments with the RDC. Although in a number of studies the filters were treated with special preparations (e.g. to render it hydrophobic) no mention about the compatibility of the filter with the chemicals used in RDC extraction experiments seems to have been made.

Despite these potential side effects the use of the RDC technique has permitted the accumulation of important information on the kinetic properties of extraction systems. So, a significant and noteworthy example is provided by a kinetic study [19] of salicylic acid extraction by liquid hydrocarbons. In this latter work, it was

---

[1]A confusion is sometimes made in which this layer is identified with the hypothetical Nernst stagnant layer appearing in the classic film theory[9]. In fact, the diffusion layer appearing in the Levich theory is not "quiescent". Moreover, the concentration profile is not linear in that layer, even if the diffusion layer thickness δ is defined by eq 3.5 as in the stagnant layer model, and the diffusion layer thickness given by eq 3.22 depends on the diffusion coefficient of the species, which is not the case in the film model.

observed that the rate constants in both directions were significantly less for interfaces between water and even-numbered hydrocarbons than those formed with odd-numbered hydrocarbons. It was concluded that a more structured interfacial region in the former systems could account for this observation. Worth of note is the fact that this amazing phenomenon has been observed also in a recent study [20] of alkane/water interfaces using the TIR-SHG technique (see end of chapter 2).

Lastly, it may be noticed that metal extraction experiments with the RDC have been performed generally at one given rotation speed ω. Sometimes the inverse of the hydrogen ion flux has been plotted against $\omega^{-1/2}$ and a linear correlation has been found [38, 40, 42], but it has been pointed out [42] that this type of Koutecki-Levich plot [41] is relevant only in the case of an interfacial mechanism. Therefore new information might be provided by studying complex extraction systems with the RDC along the rotation speed coordinate, which is an important "degree of freedom" allowed in this type of technique.

Short-time phase-contacting method

This method has been devised and used by russian workers [8, 18, 43]. It used an aqueous phase supported on a filter, which was introduced into an organic phase and transfer was allowed to occur for a short lapse of time (less than 1 s). The organic outer phase was not stirred. The amount of species extracted was followed by measuring the change in the conductivity of the aqueous phase.

This method operates fundamentally in non-stationary regime. Assuming that the organic phase was totally at rest diffusion equations could be solved analytically and the flux of matter could be expressed as a function of time. The reaction at the interface was considered as irreversible and characterized by a first-order aqueous-to-organic extraction rate constant k. It was found that the amount of matter extracted at time t after phase contact was a linear function of $t^{1/2}$ with an intercept proportional to D/k, with D the diffusion coefficient of the species in the aqueous phase.

An interesting feature of this technique was that the system operated in a regime mainly controlled by the chemical reaction because, for very short contact times, diffusion is not limiting [25] and then the effect of the kinetics is highlighted at the very beginning of the transfer process. This result was in contrast with other methods in which the stationary regime imposes a difficult separation between the diffusive and kinetic contributions.

However, it has been underlined [27] that this technique, in which the amount of matter is followed by a conductimetric method, cannot be used for concentrated solutions and when several species are implied in the extraction reaction.

Rotating membrane cell (RMC)

As in the case of the short-time phase-contacting method described above, this technique operates in a transient, non-stationary, regime that highlights the role of the chemical reaction. Moreover, this technique shares with the rotating diffusion cell (RDC) the capability of control of the hydrodynamics.

A first version of this method has been proposed in 1991 [21] in which a gel was used to immobilize the aqueous phase. This rotating stabilized cell (RSC) is described in Figure 3.7.

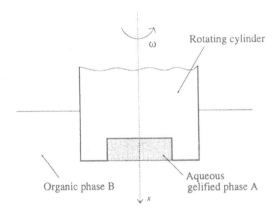

Figure 3.7: Rotating stabilized cell (RSC) technique: a polyacrylamide gel immobilizes the aqueous phase.

The RSC consists of a perspex cylinder of diameter *ca.* 1.2 cm through which a bore of *ca.* 6 mm diameter and 1 mm length was drilled. The cylinder can be mounted on a rotating-electrode spindle and can be set to rotate at a definite speed. The aqueous phase A is placed in the bore of the cylinder and it is stabilized with a gel of polyacrylamide. The presence of the gel prevents convectional motion in this phase. The pores of the gel are on the order of 0.02 μm [44]. At time $t = 0$, the cylinder, rotating at a known speed, is immersed into a definite volume of the organic solvent B. The extraction of the species is followed by sampling the organic phase at given times. At the end of the experiment the activities of the samples and that of the cylinder are measured, from which the percentage of matter extracted as a function of time can be found.

The RSC technique has been used [45] to study extraction kinetics with controlled turbulent hydrodynamics in the outer phase B. A non-phenomenological description, based on principles of turbulent hydrodynamics, was developed to interpret the data.

However, it has been found recently [23] that, in some cases, the transfer process can be blocked by some reaction involving the three constituents: the element to be extracted, the gel and the organic phase. This happened for instance with Nickel(II) and Zinc(II) ions, but not with Cobalt(II) [21], used together with HDEHP as the extractant.

In order to overcome this problem, the RSC technique has been modified by replacing the gel by a membrane: a diagram of the rotating membrane cell (RMC) [22, 23] is shown in Figure 3.8.

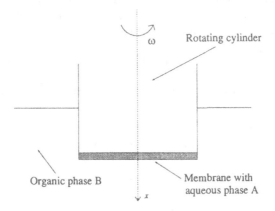

Figure 3.8: Rotating membrane cell (RMC). One of the phases is supported by a membrane (of diameter $\simeq$ 8 mm) which is glued on a perspex cylinder

The membrane was a Millipore hydrophilic PVDF membrane (HVLP type). According to the manufacturer, this type of membrane exhibits good chemical compatibility with acid solutions and aliphatic solvents and its average pore size is 0.45 μm. It was glued on the base of a cylinder made of perspex, using a polyurethanne mastic which was selected for its excellent chemical resistance to organic solvents. The diameter of the membrane was about 0.8 cm and its thickness was ca. 120 μm, giving a volume of about 4 μL. The species to be extracted was taken in radiolabelled form and the extracted amount of matter was followed in the course of time by counting the radioactivity of samples taken from organic phase.

Expressions have been given for the proportion of matter extracted as a function of time, $P(t)$, in the case of pseudo-first-order rate law for a purely interfacial reaction; the reverse stripping reaction was accounted for in this treatment. An expression was derived [21] for the case of observation times smaller than the characteristic diffusion time within the membrane, defined by

$$\tau_A = L^2/(3D_A) \tag{3.25}$$

where L is the membrane thickness and D_A the effective diffusion coefficient in the membrane (modified by membrane tortuosity).

However, the characteristics of the membrane lead to a value for τ_A of the order of 10 s, which is quite small. Besides, the transport equations cannot be solved in closed form in the non-stationary regime produced by this experimental technique. However, the difficulty could be circumvented with the use of the average-reaction-time approach [46, 47] which can be stated in the present case as

$$P(t) \simeq 1 - \exp(-t/\tau) \tag{3.26}$$

in which τ is the mean-passage time of the species for the overall transfer process. By virtue of eq 3.26 this time is defined by

$$\tau = \int_0^\infty [1 - P(t)] \, dt \tag{3.27}$$

It was shown that the mean-passage time τ splits naturally into 3 terms as follows

$$\tau = \tau_A + \tau_f + \tau_B \tag{3.28}$$

with

$$\tau_f = L/k_f$$

$$\tau_B = \sigma L \delta_B / (K D_B)$$

in which τ_f is the characteristic time for the interfacial forward reaction, τ_A is the mean diffusion time in the membrane (eq 3.25), σ is the membrane's surface porosity and τ_B is the mean residence time of the species in the diffusion layer, which results in part from the competition between stripping (at a rate σk_r) and removal by dilution in B (at a rate D_B/δ_B).

In eq 3.28 K is the B-to-A distribution ratio (eqs 3.12,3.13) and δ_B is the diffusion layer thickness in outer phase B, given by an expression analogous to eq 3.22 [32]

$$\delta_B = 1.612 \, (\nu_B/\omega)^{1/2} \, Sc_B^{-1/3} \tag{3.29}$$

with ν_B the kinematic viscosity of B, ω the rotation speed of the cell (in rad s^{-1}) and Sc$_B$ the Schmidt number in B

$$Sc_B = \nu_B/D_B$$

It is interesting to notice that eq 3.28 can be written in a way similar to eq 3.10 as

$$k^{-1} = k_A^{-1} + k_f^{-1} + \frac{\sigma}{K}k_B^{-1} \tag{3.30}$$

in which

$$k_A = 3D_A/L$$
$$k_B = D_B/\delta_B$$

with a different dependence with respect to the porosity σ due to the different role of the membrane (separator in the RDC and support for phase A in the RMC). The resistance in B is multiplied by $\sigma < 1$ in eq 3.30 because the stripping reaction occurs here at a rate σk_r, instead of k_r when phases are in free direct contact. Moreover, eq 3.30 shows that, in the case of a purely interfacial extraction process, the mean-passage-time approach leads to a "mean" resistance for the overall process that is identical to the resistance obtained for a stationary regime.

The experimental data obtained with the RSC and RMC methods have been analyzed as follows. Assuming that the reaction takes place at the interface, the rate parameter k_f can be determined as a function of the rotation speed, by a one-parameter fit (all other parameters being determined independently). If the rate-determining step is really at the interface, then the parameter k_f should be independent of the rotation speed ω. As expected, this has been shown [23, 48] to be the case for the extraction of acetic acid by isopropyl myristate. In contrast, the extraction of zinc and nickel by HDEHP led to values for k_f that increased noticeably with ω. So this technique might constitute a way to identify true interfacial reactions. However, further experimental results are required to confirm the potential of this method.

3.4 Theoretical studies

3.4.1 Locale of the extraction reaction: interfacial vs. aqueous bulk chemical reaction

So far in this chapter it has been implicitly assumed that the extraction reaction occurs at the interface.

In fact, it has been thought for a long time that the rate-determining step in metal complexation was the formation of an intermediate metal chelate complex in the bulk aqueous phase [49]. However, the introduction of high molecular weight extractants led some workers to consider the special role of the interface: it was suggested [50] 20 years ago that in such a case interfacial kinetics could have a dominant effect relatively to homogeneous reaction rates.

Such a mechanism may be expected to occur with systems involving a very hydrophobic extractant but this situation is not met often in liquid/liquid extraction systems. Although some systems may approximately be described by a model which assumes all reaction to take place at the interface, many systems may involve a mechanism which would lie halfway between bulk phase reaction and interfacial reaction mechanisms [51]: "A more probable situation is of reaction taking place in a zone in the aqueous phase adjacent to the interface. This would picture a certain amount of extractant dissolving in the aqueous phase at the interface and then diffusing into the bulk phase, reacting on the way with the solute. The solute-extractant complex would then diffuse back into the solvent phase. This situation is more likely in the case of metals extraction than that of a reaction zone in the solvent as the latter would assume free cations transferring into an organic phase" (Abramzon and Kogan's postulate [52]). These arguments constitute the basis of the so-called mass transfer with chemical reaction (MTWCR) model. Various versions of this model have been proposed [51, 53, 54, 57] and applied primarily to copper extraction studies. The MTWCR hypothesis was supported by the observation that extractants with high extractant solubilities often extract at greater rates [51]; a correlation has also been observed between extractant partitioning kinetics and overall extraction kinetics [58].

Let us notice that a criterion has been often used to distinguish between interfacial and aqueous bulk reactions. It is based [27] on the dependence of extraction rates upon the specific area of the system, defined by eq 3.19: for an aqueous bulk reaction the extraction rate should be independent of this parameter, because the rate is governed by homogeneous chemical reactions if the transport of the extractant to the aqueous phase, and of the chelated species back to the organic, are not rate determining; on the other hand, for an interfacial reaction the extraction rate should vary in proportion to $\bar{a}$ as shown by eq 3.18. However, for reactions that would occur in a thin film close to the interface (MTWCR hypothesis) application of this criterion would wrongly point to a true interfacial reaction.

Thus, in many cases, the determination of the locale of the rate-determining step has been the subject of much debate and controversy. Significant examples may be represented by *(i)* the case of the extraction of zinc(II) by a thiocarbazone (dithizone) which has been reported to be controlled either by an interfacial mechanism in some studies or by a bulk reaction in other studies [8] and *(ii)* the extraction of copper by an hydroxyoxime (LIX65N) described by an interfacial process [55] or a MTWCR process [61]. However, in some particular cases, an interfacial mechanism has been determined: for the extraction of acetic acid by isopropyl myristate [23] consisting only of a change of solvation at the interface; for the extraction of germanium by Kelex 100S (8-hydroxyquinoline) [56] leading to a very slow process.

3.4.2 MTWCR model

The more popular version of the MTWCR model seems to have been that of Rod [57]. It has been used to describe the case of divalent metal extraction by an acidic extractant [38, 58, 59, 60, 61]. Its basic ingredients are the following.

Basic assumptions
Let HR denote the acid extractant, M^{2+} the metal ion and MR_2 the extracted complex. Then the extraction reaction is considered to proceed according to the following steps

$$\text{Step 1:} \quad \overline{HR} \rightleftharpoons HR$$
$$\text{Step 2:} \quad HR \rightleftharpoons H^+ + R^-$$
$$\text{Step 3:} \quad M^{2+} + R^- \overset{k_R}{\rightleftharpoons} MR^+$$
$$\text{Step 4:} \quad MR^+ + R^- \rightleftharpoons MR_2$$
$$\text{Step 5:} \quad MR_2 \rightleftharpoons \overline{MR_2}$$

where the bar notation indicates an organic phase species and its absence refers to an aqueous phase species.

In the above reactions the first step describes the fast partitioning of the extractant in the aqueous phase followed by fast dissociation of the extractant in step two; the third and fourth steps describe the formation of the metal complexes; the last step involves the rapid partitioning of the metal complex into the organic phase.

Some acid extractants, such as D2EHPA, are strongly dimerized in the organic phase and consequently step 1 should be replaced by $\overline{HR_2} \rightleftharpoons 2\,HR$ [62] in that case; it has been also assumed sometimes [61] that in this case the complexation reaction in the aqueous phase involved the species HR_2^-.

We now present the main equations and results given in ref. [57]. The ingredients of this MTWCR model do not seem to have been reviewed so far.

The MTWCR model presented below was first proposed by Rod [57]. Due to the complexity of the problem, the following assumptions were adopted

- The model was developed in the framework of the two-film model [9]. It was therefore assumed that the species diffuse in a quiescent layer in the aqueous phase with chemical reactions described by steps 2, 3 and 4.

- The system operates in a stationary regime.

- All reactions in the bulk phase are fast and therefore equilibrated.

- Step 3 is the rate-determining step, with forward rate constant k_R. Steps 2 and 4 are very fast and the related local concentrations are in equilibrium.

- The net flux of the active species R in all forms is zero at each location in the layer.

- Concentrations of the species R^- and MR^+ are low compared to those of HR and H^+; the fluxes of the species R^- and MR^+ are much smaller than those of the other species.

Moreover, it was implicitly assumed that

- The system is thermodynamically ideal: activity coefficient effects are not considered.

- The interfacial concentrations of the extractant HR and the metal complex MR_2 are equilibrated. The finite fluxes of these species at the interface do not modify the organic-to-aqueous interfacial concentration ratios.

- The flux of an ionic species k is written as

$$j_k = -D_k \frac{dC_k}{dx} \tag{3.31}$$

i.e. the purely diffusive flux given by Fick's law. Electric diffusive coupling [63, 64], arising from electrostatic interactions between ions, was not taken into account. Although this effect may be neglected in presence of a support electrolyte which screens out the electrostatic interactions, it might constitute a crude approximation in absence of support electrolyte.

Typical concentration profiles [59] for the MTWCR model are shown in Figure 3.9.

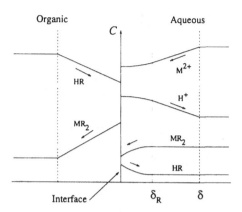

Figure 3.9: Concentration profiles in the MTWCR model (quiescent films in both phases and reaction zone denoted by dashed lines and dotted line, respectively).

Basic equations

The transport continuity equation for the species HR in the aqueous quiescent layer gives

$$\frac{\partial C_{HR}}{\partial t} \equiv 0 = D_{HR}\frac{d^2C_{HR}}{dx^2} + \sigma_{HR} \tag{3.32}$$

in which σ_{HR} denotes the chemical reaction rate term for the production of HR (in step 2), which is

$$\sigma_{HR} = -\sigma_{R-} = -k_R C_{M^{2+}}C_{R-} + k'_R C_{MR+} \tag{3.33}$$

with k'_R the reverse rate constant of step 2 which is related to the equilibrium association constant of step 3 by

$$K_3 = k_R/k'_R = C_{MR+}/(C_{M^{2+}}C_{R-}) \tag{3.34}$$

The other equilibrium constants are defined by

$$K_2 = C_{HR}/(C_{H+}C_{R-}) \tag{3.35}$$

$$K_4 = C_{MR_2}/(C_{MR+}C_{R-}) \tag{3.36}$$

and the partition coefficients for steps 1 and 5 are defined as the organic-to-aqueous distribution ratios

$$K_1 = \bar{C}_{HR,i}/C_{HR,i} \tag{3.37}$$

$$K_5 = \bar{C}_{MR_2,i}/C_{MR_2,i} \tag{3.38}$$

in which the notation $\bar{C}_{S,i}$ denotes the concentration of a species S on the organic side of the interface.

Then the equilibrium constant for the overall reaction

$$2\,\overline{HR} + M^{2+} \rightleftharpoons \overline{MR_2} + 2\,H^+$$

is

$$K^* = \frac{K_5}{K_1^2}\,K_{eq} \tag{3.39}$$

with

$$K_{eq} = K_3 K_4/K_2^2 \tag{3.40}$$

With these notations eqs 3.32 and 3.33 can be rewritten as

$$D_{HR}\frac{d^2 C_{HR}}{dx^2} = \frac{k_R}{K_2 C_{H^+}}\left[C_{HR}C_{M^{2+}} - C_{H^+}^2 C_{MR_2}/(K_{eq}C_{HR})\right] \tag{3.41}$$

Besides, it was stated that under the above assumptions the following relations held

$$D_{H^+}\frac{dC_{H^+}}{dx} + 2D_{M^{2+}}\frac{dC_{M^{2+}}}{dx} = 0 \tag{3.42}$$

$$D_{HR}\frac{dC_{HR}}{dx} + 2\frac{dC_{MR_2}}{dx} = 0 \tag{3.43}$$

$$-D_{HR}\frac{dC_{HR}}{dx} + 2D_{M^{2+}}\frac{dC_{M^{2+}}}{dx} = j_{HR,i} \tag{3.44}$$

However these relations were given without explicit proof and we may try now to justify them.

- Eq 3.42 may be shown as follows. The combination of steps 2, 3 and 4 shows that the reaction of one metal ion M^{2+} is accompanied by the release of 2 protons H^+. Then, if the reaction is fast, the two fluxes should compensate locally. This result can be shown by assuming a quasi-stationary state for the local concentrations of R^- and MR^+, which entails

$$\sigma_{R^-}^{(t)} = 0 = \sigma_{H^+}^{(2)} + \sigma_{M^{2+}}^{(3)} + \sigma_{R^-}^{(4)}$$

meaning that the total production rate of R^- comes from steps 2, 3 and 4, and

$$\sigma_{MR^+}^{(t)} = 0 = -\sigma_{M^{2+}}^{(3)} + \sigma_{R^-}^{(4)}$$

Combining these last two equations yields

$$\sigma_{H^+}^{(2)} + 2\sigma_{M^{2+}}^{(3)} = 0$$

Then, using this relation and the two continuity equations for H^+ and M^{2+}

$$\frac{\partial C_{H^+}}{\partial t} = 0 = -\frac{d\,j_{H^+}}{dx} + \sigma_{H^+}^{(2)}$$

$$\frac{\partial C_{M^{2+}}}{\partial t} = 0 = -\frac{d\,j_{M^{2+}}}{dx} + \sigma_{M^{2+}}^{(3)}$$

gives

$$\frac{d}{dx}\left(j_{H^+} + 2j_{M^{2+}}\right) = 0$$

and since there is no ionic flux to the organic phase at the interface (boundary condition, eq 3.47 below) one obtains

$$j_{H^+} + 2j_{M^{2+}} = 0 \tag{3.45}$$

which, by virtue of assumption 3.31, gives eq 3.42.

- Eq 3.43 results from the above assumption that the net diffusive flux of species R is zero and that the fluxes of species R^- and MR^+ can be neglected.

- The conservation of the flux of species H (*i.e.* H^+ and HR) leads to

$$j_{H^+} + j_{HR} = j_{HR,i}$$

because the flux of H originates only from the interfacial flux density of HR, denoted by $j_{HR,i}$. Then, by virtue of this equation and eqs 3.31, 3.45 we obtain eq 3.44.

The last relation given in ref. [57] was

$$K_{eq} = C_{MR_2,bulk}C_{H^+,bulk}/(C_{HR,bulk}^2 C_{M^{2+},bulk}) \tag{3.46}$$

It results from the assumption that the chemical reactions are equilibrated in the bulk aqueous phase (steps 2+3+4). In this equation the aqueous bulk concentrations of MR_2 and HR are not zero, as might be deduced from the last relation of eq 3.43

and the last boundary condition of eqs 3.48, because of transfer of these species during the first stage of the extraction process, before establishment of the stationary regime.

The boundary conditions for this problem were written as

$$\text{At } x = 0: \quad C_{HR} = C_{HR,i}; \quad C_{MR_2} = C_{MR_2,i}; \quad dC_{M^{2+}}/dx = 0 \quad (3.47)$$

$$\text{At } x = \delta: \quad C_{H^+} = C_{H^+,bulk}; \quad C_{M^{2+}} = C_{M^{2+},bulk}; \quad dC_{HR}/dx = 0 \quad (3.48)$$

with δ the thickness of the aqueous film. In eq 3.47 the condition on the concentration gradient of M^{2+} expresses the fact that this ion cannot be transferred in this form to the organic phase. In eq 3.48 the similar condition on the flux of HR means that it is assumed that passage of HR to the bulk aqueous phase is not allowed.

Results

The interfacial flux of the species M, giving the extraction flux of M, is given by the relation

$$j_{MR_2,i} = -\frac{1}{2} j_{HR,i} \quad (3.49)$$

where $j_{HR,i}$ appears in eq 3.44.

The solution to eqs 3.41-3.48 cannot be obtained in closed form. Approximate solutions were obtained in limiting cases. In particular, an expression was given [65] in the case of an "instantaneous" reaction, indicated by the dimensionless parameter

$$M = k_R D_{HR} C_{M^{2+},bulk}/(k_{HR,aq}^2 K_2 C_{H^+,bulk}) \quad (3.50)$$

being much greater than unity (the criterion given in ref. [57] was $M \to \infty$). In this equation k_{HR} represents the mass transfer coefficient in the aqueous layer of thickness δ

$$k_{HR,aq} = D_{HR}/\delta \quad (3.51)$$

In that case eq 3.41 was approximated by the following equation

$$D_{HR}\frac{d^2 C_{HR}}{dx^2} = \frac{k_R}{K_2 C_{H^+,i}} \left[C_{M^{2+},i} - C_{H^+,i}^2 C_{MR_2,i}/(K_{eq} C_{HR,i}^2) \right] C_{HR} \quad (3.52)$$

after correction of a misprint. Derivation of eq 3.52 was not justified in ref. [57]. However this relation may be obtained by linearization of eq 3.41 in which the concentration C_{HR} is factorized and the remaining concentrations are replaced by their interfacial values.

Then eq 3.52, together with the boundary conditions, was solved, which led to [65]

$$j_{HR,i}^2 = \theta_1 \frac{C_{M^{2+},i}}{C_{H^+,i}} \left(1 - \frac{C_{H^+,i}^2 \bar{C}_{MR_2,i}}{K^* \bar{C}_{HR,i}^2 C_{M^{2+},i}}\right) (\bar{C}_{HR,i}^2 - \bar{C}_{HR,*}^2) \qquad (3.53)$$

in which the notation $\bar{C}$ denotes the concentration of a species on the organic side of the interface (MR_2 and HR concentrations are not continuous at the interface).

$$\bar{C}_{HR,*} = -\frac{\theta_2 C}{4} + \left[\left(\frac{\theta_2 C}{4}\right)^2 + C(\bar{C}_{MR_2,i} + \theta_2 \bar{C}_{HR,i})\right]^{1/2} \qquad (3.54)$$

with

$$\theta_1 = k_R D_{HR}/(K_1^2 K_2) \qquad (3.55)$$

$$\theta_2 = D_{HR} K_5/(2 D_{MR_2} K_1) \qquad (3.56)$$

$$C = C_{H^+,bulk}^2/(K^* C_{M^{2+},bulk}) \qquad (3.57)$$

and interfacial concentrations were given by

$$\bar{C}_{HR,i} = \bar{C}_{HR,bulk} - j_{HR,i}/k_{HR,org} \qquad (3.58)$$

$$\bar{C}_{MR_2,i} = \bar{C}_{MR_2,bulk} + \frac{1}{2} j_{HR,i}/k_{MR_2,org} \qquad (3.59)$$

$$C_{M^{2+},i} = C_{M^{2+},bulk} - j_{HR,i}/k_{M^{2+},aq} \qquad (3.60)$$

$$C_{H^+,i} = C_{H^+,bulk} + \frac{1}{2} j_{HR,i}/k_{H^+,aq} \qquad (3.61)$$

where the parameters $k_{X,aq}$ and $k_{X,org}$ denote mass transfer coefficients defined as in eq 3.51.

In refs. [59, 61, 68] the last two equations 3.60, 3.61 were written in a different way as

$$C_{M^{2+},i} = C_{M^{2+},bulk} - j_{HR,i}/(2k_{M^{2+},aq}) + [D_{HR}/(2D_{H^+}K_1)](\bar{C}_{HR,i} - \bar{C}_{HR,*}) \quad (3.62)$$

$$C_{H^+,i} = C_{H^+,bulk} + j_{HR,i}/k_{H^+,aq} - [D_{HR}/(2D_{H^+}K_1)](\bar{C}_{HR,i} - \bar{C}_{HR,*}) \quad (3.63)$$

Therefore the model predicts an overall extraction rate (eqs 3.53-3.61) which shows a complex non-linear dependence on concentrations.

The accuracy of these approximate results has been examined [57] by comparing them with results from a numerical analysis of eqs 3.41-3.48, for some parameter values: it was found that the mean deviation of the approximate result from the numerical one was about 3 %, with a maximum deviation of the order of 10%.

The thickness of the reaction zone has been given [59] as

$$\delta_R \simeq 3\frac{D_{HR}}{K_1}\left(\frac{\bar{C}_{HR}}{j_{HR,i}} - \frac{1}{k_{HR,org}}\right) \tag{3.64}$$

Let us notice lastly that, in refs. [59, 68], it was stated that eq 3.53 could be obtained by Astarita's method [66], used previously in a pioneering MTWCR model [51]. The use of Astarita's method within the MTWCR model of Rod has been explained in a PhD thesis [67].

Application of the MTWCR model

First, in at least two studies [38, 58] various interfacial mechanisms have been found inadequate for describing the experimental data, which led the authors to use the MTWCR model.

The MTWCR model introduced by Rod has been used to interpret mainly experiments performed with the RDC of Albery because this technique allows an estimation of the diffusion layer thickness. However, since the latter depends on the species considered through its diffusion coefficient (cf. eq 3.22), it is not clear which value was taken for the film thickness, which is supposed to be unique for a given phase in the film model, for all the species involved (cf. Figure 9).

The experiments fitted with the model included the extraction of copper by hydroxyoximes (LIX64N, P5000,...) [59, 61, 65] and dialkylphosphoric acids [61], of zinc, cobalt and nickel by HDEHP [38, 60, 62] or HEHEHP [58]. In most cases the model could be fitted to the experimental results, by the adjustment of the parameters θ_1, θ_2 and K^*, defined by eqs 3.39, 3.55, 3.56. The sensitivity of the model to the grouped parameter θ_2 has been found to be low [65, 59, 61].

However, in some cases (e.g. [38]) the model was not successful. This happened in the case of the extraction of zinc and nickel with HDEHP using the RDC technique, and this was interpreted as follows: in the case of zinc the transfer was found to be controlled by mass transfer alone because the chemical reaction was too fast to limit the kinetics; in the case of nickel the reaction was too slow and much extractant was partitioning to the aqueous phase without complexing the metal cation, thus making it impossible to use the MTWCR model of Rod [57] presented above.

When the model could be fitted to experimental data values for the rate constant k_R could be estimated from the adjustment of θ_1 value (eq 3.55) and measurements or estimations of the parameters D_{HR}, K_1 and K_2. These k_R values were regarded as representative of the complex formation rate constants for water replacement on the ion by water soluble extractant anions. So, values of 26 m^3 (kmol s)$^{-1}$ and 0.00468 m^3 (kmol s)$^{-1}$ have been found [58] for the complexation of cobalt and nickel by HEHEHP, and $(2-9)\times 10^{10}$ m^3 (kmol s)$^{-1}$ for the reaction of copper with the hydroxyoxime P50 [61]. Besides, in this latter study, the reaction zone thickness δ_R (eq 3.64) was found to be 0.7 μm for the Cu^{2+}/HDEHP/heptane system and 13 Å for the Cu^{2+}/P50/heptane system, giving in the latter case a reaction zone of microscopic dimension, indicating probably an interfacial reaction in agreement with a very low solubility of P50 in the aqueous phase and with a previous work of Albery *et al.* [42].

3.4.3 Numerical studies

A few molecular dynamics computer simulations have been undertaken in the last decade. The interest of this type of calculation lies in its ability to study systems at a microscopic level [70] and some striking new results are now beginning to emerge from such studies.

A model of immiscible Lennard-Jones atomic solvents has been used to study the adsorption of a diatomic solute [71]. Subsequently, studies of solute transfer have been performed for atoms interacting through Lennard-Jones potentials [69] and an ion crossing an interface between a polar and a nonpolar liquid [72]. In both cases the potential of mean force experienced by the solute was computed; the results of the simulation were compared with the result from the transition state theory (TST) in the first case, and with the result from a diffusion equation in the second case. The latter comparison has led to the conclusion that the rate calculated from the molecular dynamics trajectories agreed with the rate calculated using the diffusion equation, provided the mean-force potential and the diffusion coefficient were obtained from the microscopic model.

A more realistic system has been approached recently [73], concerning the transfer of small ions (such as Cl^-, Na^+,...) across the water/1,2-dichloroethane interface. No artificial constraints were necessary to ensure solvent immiscibility, which was the result of the hydrogen-bonding forces in the water. So, the contact of the two liquids gave a sharp interface, deformed by internal capillary waves. Transfer of the ion was helped by application of an external electric field, in the range 0.1-0.3 V/Å, of a magnitude comparable to the fields commonly used in voltammetric studies. In another similar work [74], some results have been reported for the case of no field applied. The first main result was that the ion solvation state should be included as a new coordinate when describing the transfer by a barrier crossing; thus the transfer can be viewed as a transport process on a two-dimensional free energy surface (the distance from the interface and the solvation coordinate). Next, examination of the ion trajectories showed the following phenomena: the transfer from the water to the organic phase showed clearly that, on the time scale of the simulation, the ion dragged its hydration shell into the organic phase, with a "chain" of water molecules connecting the hydrated ion to the water phase; reversly, the transfer from the organic phase to the water was found to be effected by the interaction of the ion with a water "finger" extending out of the water phase into the organic phase. These "fingers" were associated with the interfacial capillary waves; the amplitude of these water protrusions was on average 6 Å long and could be as long as 10 Å, so yielding an appreciable interface roughness. Moreover, the transfer was examined at the energy level: for an ion starting in the dichloroethane, the rate-limiting step was determined to be its ability to find a water molecule which, although attached to the water phase, is capable of hydrating the ion; this process was observed to be facilitated by the presence of water "fingers" whose head water molecule is more loosely connected to the other water molecules. In contrast with a previous work [69] the switching of the first solvation shell was found to be fast and not activated, and the barrier to transfer was suggested to have an entropic part associated with the need for matching the ion with a water "finger" plus an energetic part related to the breaking of some hydrogen bonds for the creation of such a "finger". However, it was pointed out that the transfer of such small ions, exhibiting large energies of transfer, could be facilitated by the formation of ion pairs such as NaCl.

Recently, various real extraction systems have also been studied using an Amber software [75].

Bibliography

[1] *Principles and Practices of Solvent Extraction*, Eds. J. Rydberg, C. Musikas and G.R. Choppin, Dekker, New York, 1992.

[2] H.H.J. Girault and D.J. Schiffrin in *Electroanalytical Chemistry*, Ed. A.J. Bard, Dekker, New York, 1989.

[3] *The Interface Structure and Electrochemical Processes at the Boundary between Two Immiscible Liquids*, Ed. V.E. Kazarinov, Springer, Berlin, 1987.

[4] N. Lakshminarayanaiah *Equations of Membrane Biophysics*, Academic Press, New York, 1984.

[5] *Membrane Transport in Biology*, Eds. G. Giebisch, D.C. Tosteson and H.H. Ussing, Springer, Berlin, 1978.

[6] B.H. Honig, W.L. Hubbell and R.F. Flewelling *Ann. Rev. Biophys. Biophys. Chem.* 15 (1986) 163.

[7] E.V. Sitzmann and K.B. Eisenthal *J. Phys. Chem.* 92 (1988) 4579; *ibid. J. Chem. Phys.* 90 (1989) 2831.

[8] P.R. Danesi and R. Chiarizia *CRC Crit. Rev. Anal. Chem.* 10 (1980) 1-126.

[9] W.G. Whitman *Chem. Met. Eng.* 29 (1923) 147; Ind. Eng. Chem. 16 (1924) 1215.

[10] P.R. Danesi, G.F. Vandegrift, E.P. Horwitz and R. Chiarizia *J. Phys. Chem.* 84 (1980) 3582.

[11] P.R. Danesi *Solvent Extr. Ion Exch.* 2 (1984) 29.

[12] J.B. Lewis *Chem. Eng. Sci.* 3 (1954) 218.

263

[13] P.R. Danesi, C. Cianetti, E.P. Horwitz and H. Diamond *Sep. Sci. Technol.* 17 (1982) 961.

[14] W. Nitsch and K. Hillekamp *Chem. Ztg.* 96 (1972) 254.

[15] J. Rydberg, H. Reinhardt and J.O. Liljenzin *Ion Exch. Solvent Extr.* 3 (1973) 111.

[16] W. Nitsch *Dechema Monogr.* 55 (1965) 143.

[17] W.J. Albery, J.F. Burke, E.B. Leffler and J. Hadgraft *J. Chem. Soc. Faraday Trans.* 72 (1976) 1618.

[18] V.V. Tarasov, N.F. Kizim and G.A. Yagodin *Russ. J. Phys. Chem.* 45 (1971) 1425.

[19] R.H. Guy, T.R. Aquino and D.H. Honda *J. Phys. Chem.* 86 (1982) 280.

[20] J.C. Conboy, J.L. Daschbach and G.L. Richmond *J. Phys. Chem.* 98 (1994) 9688.

[21] J.P. Simonin, P. Turq and C. Musikas *J. Chem. Soc. Faraday Trans.* 87 (1991) 2715.

[22] J.P. Simonin *Solvent Extr. Ion Exch.* 14 (1996) 889.

[23] J.P. Simonin and J. Weill *Solvent Extr. Ion Exch.* 16(6) (1998), to be published.

[24] G.J. Hanna and R.D. Noble *Chem. Rev.* 85 (1985) 583.

[25] D.A. Frank-Kamenetskii *Diffusion and Heat Transfer in Chemical Kinetics*, Plenum Press, New York, 1969.

[26] M.A. Hughes and V. Rod *Hydrometallurgy* 12 (1984) 267.

[27] P.R. Danesi in ref. [1], p. 157.

[28] P.V. Danckwerts *Ind. Eng. Chem.* 43 (1951) 1460.

[29] H.R.C. Pratt in *Handbook of Solvent Extraction*, Eds. T.C. Lo, M.H.I. Baird and C. Hanson, Wiley, New York, 1983, p. 91-123.

[30] W.J. Albery, A.M. Couper, J. Hadgraft and C. Ryan *J. Chem. Soc. Faraday Trans. I* 70 (1974) 1124.

[31] T. von Karman *Z. Angew. Math. Mech.* 1 (1921) 244.

[32] V.G. Levich *Physicochemical Hydrodynamics*, Prentice Hall, Englewood Cliffs, NJ, 1962.

[33] J. Newman *J. Phys. Chem.* 70 (1966) 1327.

[34] W.H. Smyrl and J. Newman *J. Electrochem. Soc.* 118 (1971) 1079.

[35] N.H. Sagert, M.J. Quinn and R.S. Dixon *Can. J. Chem.* 59 (1981) 1096.

[36] R.H. Guy, T.R. Aquino and D.H. Honda *J. Phys. Chem.* 86 (1982) 280.

[37] D.B. Dreisinger and W.C. Cooper *Solvent Extr. Ion Exch.* 4 (1986) 135.

[38] D.B. Dreisinger and W.C. Cooper *Solvent Extr. Ion Exch.* 7 (1989) 335.

[39] D.W.J. McLean and D.B. Dreisinger *Hydrometallurgy* 33 (1993) 107.

[40] M.A. Hughes and R.K. Biswas *Hydrometallurgy* 32 (1993) 209.

[41] J. Koutecky and V.G. Levich *Zh. Fiz. Khim.* 32 (1956) 1565.

[42] W.J. Albery and R.A. Choudhery *J. Phys. Chem.* 92 (1988) 1142.

[43] V.V. Tarasov and G.A. Yagodin in *Ion Exchange and Solvent Extraction*, Eds. J.A. Marinski and Y. Marcus, vol. 10, chapter 4, p. 141, 1988.

[44] A.T. Andrews *Electrophoresis: theory, techniques, and biochemical and clinical applications*, 2nd ed., Clarendon Press, Oxford, 1986.

[45] J.P. Simonin *Solvent Extr. Ion Exch.* 15 (1997) 483.

[46] A. Szabo, K. Schulten and Z. Schulten *J. Chem. Phys.* 72 (1980) 4350.

[47] J.P. Simonin and M. Moreau *Mol. Phys.* 70 (1990) 265.

[48] J.P. Simonin *Solvent Extr. Ion Exch.* 13 (1995) 941.

[49] H. Freiser *Chem. Rev.* 88 (1988) 611.

[50] D. Flett *Acc. Chem. Res.* 10 (1977) 99.

[51] C. Hanson, M.A. Hughes and J.G. Marsland *Proceedings of ISEC'74*, Society of Chemical Industry, London, vol. 3, p. 2401, 1974.

[52] A.A. Abramzon and N.A. Kogan *J. Appl. Chem. (USSR)* 36 (1963) 1949.

[53] K. Kondo, S. Takahashi, T. Tsuneyuki and F. Nakashio *J. Chem. Eng. Japan* 11 (1978) 193.

[54] R.J. Whewell, M.A. Hughes and C. Hanson *Proceedings of ISEC'77*, Can. Inst. Min. Metal., vol. 21, p. 185, 1979.

[55] H.Y. Lee, S.G. Kim and J.K. Oh *Hydrometallurgy* 34 (1994) 293.

[56] C. Bouvier *Contribution l'tude des effets de sels sur la cintique de transfert de phase en extraction liquide-liquide; application l'hydromtallurgie*, Ph.D. Thesis, ESPCI Paris, Universit Paris VI, France, 1997.

[57] V. Rod *Chem. Eng. J.* 20 (1981) 131.

[58] D.B. Dreisinger and W.C. Cooper *Solvent Extr. Ion Exch.* 4 (1986) 317.

[59] M.A. Hughes and V. Rod *Faraday Disc. Chem. Soc.* 77 (1984) 75.

[60] H.V. Patel *The Kinetics of Liquid-Liquid Extraction of Metal in a Rotating Diffusion Cell*, Ph.D. Thesis, Department of Chemical Engineering, University of Bradford, UK, 1988.

[61] M.A. Hughes and P.K. Kuipa *Ind. Eng. Chem. Res.* 35 (1996) 1976.

[62] M.A. Hughes and T. Zhu *Hydrometallurgy* 13 (1985) 249.

[63] R. Mills, A. Perera, J.P. Simonin, L. Orcil and P. Turq *J. Phys. Chem.* 89 (1985) 2722.

[64] J.P. Simonin, J.F. Gaillard, P. Turq and E. Soualhia *J. Phys. Chem.* 92 (1988) 1696.

[65] V. Rod, L. Strnadova, V. Hancil and Z. Sir *Chem. Eng. J.* 21 (1981) 187.

[66] G. Astarita *Mass Transfer with Chemical Reaction*, Elsevier, Amsterdam, 1967.

[67] P.K. Kuipa, PhD Thesis, *The Kinetics of Metal Extraction relating to Extractant Interfacial Activity and Aqueous Phase Solubility*, University of Bradford, 1995.

[68] M.A. Hughes and V. Rod *Proceedings of ISEC'83*, ACS 14, 1983.

[69] M. Hayoun, M. Meyer and P. Turq *J. Phys. Chem.* 98 (1994) 6626.

[70] I. Benjamin *Chem. Rev.* 96 (1996) 1449.

[71] M. Mareschal in *Chemical Reactivity in Liquids*, M. Moreau and P. Turq Eds., Plenum Press, New York, 1987.

[72] I. Benjamin *J. Chem. Phys.* 96 (1992) 577.

[73] K.J. Schweighofer and I. Benjamin *J. Phys. Chem.* 99 (1995) 9974.

[74] I. Benjamin *Science* 261 (1993) 1558.

[75] A. Varnek, L. Troxler and G. Wipff *Chem. Eur. J.* 3 (1997) 552; M. Lauterbach, E. Engler, N. Muzet, L. Troxler and G. Wipff *J. Phys. Chem.* 102 (1998) 245; and references cited in these articles.

Chapter 4

Electrokinetic phenomena

4.1 Introduction

Electrokinetic phenomena designate the transport phenomena involving electrolytes near charged interfaces.

Obviously the boundary conditions have a considerable influence on the observed transport processes.

If the electrolyte flow moves perpendicularly to the interface, it will be stopped by it, except if some particular event such as an electrochemical reaction occurs, giving a source (or a well) of solute. Moreover, in this situation, the solvent, which is in most liquids uncompressible, will have to escape (or to arrive) in another direction and then to move in a direction parallel to the interface.

The most common situation concerning the motion of electrolytes near interfaces is then that concerning a motion parallel to the interface. This will involve not only the solute flow, but also the solvent flow.

Since most interfaces in contact with a liquid bear a superficial charge, the combination between hydrodynamic and electrostatic conditions will be the key of the observed processes, which are known as eletrokinetic phenomena.

In this chapter, we recall briefly some features of ionic transport in solutions [1]. Since the basic concepts of electrostatics and hydrodynamics have been presented before, we will directly present their application to electrokinetic phenomena after this first presentation.

4.2   Transport in ionic liquids

4.2.1   Limiting velocity

Under the action of forces due to other particles j, or to external fields, a given particle i is accelerated as:

$$m_i \frac{d\mathbf{v_i}}{dt} = \sum_j \mathbf{F_{ji}}$$

Due to the solvent, a friction force $-\zeta_i \mathbf{v_i}$ appears. For a particle submitted to the action of an external electrical field, we have:

$$m_i \frac{d\mathbf{v_i}}{dt} = Z_i e\mathbf{E} - \zeta_i \mathbf{v_i}$$

The particle takes a limiting velocity

$$\mathbf{v_i}^{lim} = \frac{Z_i e\mathbf{E}}{\zeta_i}$$

This velocity is reached in an exponential manner

$$\mathbf{v_i}(t) = \frac{Z_i e\mathbf{E}}{\zeta_i} (1 - \exp(-t/\tau_i))$$

where the characteristic time τ_i is given by

$$\tau_i = \frac{m_i}{\zeta_i}$$

4.2.2   Diffusion

External forces on particles can be of origins other than electrostatic. Let us consider, as another example, the case of diffusion forces.

Diffusion forces derive from macroscopic chemical potentials. In the simplest case, of an ideal solution we have

$$\mu_i = \mu_i^{ref} + kT \ln c_i$$

$$\mathbf{F_i} = -\nabla\mu_i = -kT \frac{\nabla c_i}{c_i}$$

The velocity taken under the action of the diffusion force by particle i is then:

$$\mathbf{v_i} = \frac{kT}{\zeta_i} \frac{\nabla c_i}{c_i}$$

4.2.3 Flow of particles

We now consider the flow of particles generated by diffusion.

This flow will be related to the local concentration c_i of species i

$$c_i = \frac{n_i}{V}$$

In a cylinder of section S and length l, let we denote by $\mathbf{J}_i$ the number of particles crossing S by time unit.

if $S = 1$ surface unit, then $V = |\mathbf{v}_i|$ and $|\mathbf{J}_i| = n_i = c_i V$ and $\mathbf{J}_i = c_i \mathbf{v}_i$

Diffusion flow

In this paragraph the basic relation between diffusional flow and concentration gradient (Fick's law), is presented.

Introducing the diffusion velocity in the flow gives

$$\mathbf{J}_i = c_i \mathbf{v}_i = -\frac{kT}{\zeta_i} \nabla c_i$$

We define the diffusion coefficient as

$$D_i = \frac{kT}{\zeta_i}$$

And get the first Fick's law

$$\mathbf{J}_i = -D_i \nabla c_i$$

Charged particles

The same evaluation can be done for charged particles and will give the relation between electrical and diffusional transport known as Nernst'Einstein relation. The expression of the charge flow will also give Ohm's law for ionic transport in an electrolyte.

The velocity is due to the action of the electric field

$$\mathbf{v}_i = Z_i e \frac{\mathbf{E}}{\zeta_i} = u_i \mathbf{E}$$

The relation between the diffusional and electrical mobility is then

$$\frac{Z_i e}{\zeta_i} = u_i = Z_i e \frac{D_i}{kT}$$

or

$$\frac{D_i}{u_i} = \frac{kT}{Z_i e}$$

which is Nernst Einstein's relation.

We have for the flow of charged particles

$$\mathbf{J}_i = c_i \mathbf{v}_i = Z_i \frac{e}{\zeta_i} c_i \mathbf{E}$$

and for the corresponding flow of charges

$$\mathbf{J}_i^{el} = Z_i e \mathbf{J}_i = Z_i^2 e^2 \frac{c_i}{\zeta_i} \mathbf{E}$$

$$\mathbf{J}_T^{el} = \sum_i \mathbf{J}_i^{el} = \left(\sum_i Z_i^2 e^2 \frac{c_i}{\zeta_i} \right) \mathbf{E} = \gamma \mathbf{E}$$

which is simply Ohm's law

Inertial time

Einstein's relation allows for the explicit expression of the inertial time τ_i as a function of the diffusion coefficient D_i

$$\frac{kT}{\zeta_i} = D_i \qquad\qquad \tau_i = \frac{m_i}{\zeta_i} = \frac{D_i m_i}{kT}$$

Its order of magnitude for most ions ($m_i \simeq 100$ gram per mol, $D_i \simeq 10^{-5}$ cm^{2}s^{-1}) is a picosecond:

$$\tau_i \simeq 10^{-12} s$$

Equivalent conductivity

For the sake of simplicity, let us consider a simple symmetrical 1-1 electrolyte such as NaCl.

The charges and concentrations are

$$Z_1 = 1, \quad Z_2 = -Z_1 = -1$$

$$c_1 = c_2 = c$$

and the conductivity

$$\gamma = \frac{e^2}{kT}(D_+ + D_-)c$$

$$\frac{\gamma}{c} = \frac{e^2}{kT}(D_+ + D_-) = \Lambda$$

is the equivalent conductivity

For practical systems, as will be seen in the part on transport with MSA (mean spherical approximation), the equivalent conductivity varies with concentration.

Mutual diffusion, junction potentials

The notion of charge flow can also be applied to a situation without external electric field. Besides the static electroneutrality condition

$$\sum_i Z_i e c_i = 0$$

which represents the local electroneutrality we have to introduce a dynamic electroneutrality condition, which prevents any macroscopic charge separation. This condition is introduced at a semi-macroscopic level and expresses simply the fact that the flows of charges have to be balanced to keep the local electroneutrality.

$$\sum_i Z_i e \mathbf{J}_i = 0$$

This condition can be satisfied only by introducing a local internal electric field, which prevents any dynamic charge separation

$$\mathbf{J}_i = -D_i \nabla c_i + Z_i e \frac{D_i}{kT} c_i \mathbf{E}_{int}$$

The dynamic electroneutrality condition can then be satisfied only by the presence of a local internal field

$$\mathbf{E}_{int} = \frac{kT}{e} \frac{\sum_i Z_i D_i \nabla c_i}{\sum_j Z_j^2 D_j c_j}$$

This expression is known as Henderson's formula.

Electrolyte diffusion

As for the electric conductivity, the obtained results are very simple for 1-1 electrolytes

$$\mathbf{E}_{int} = \frac{kT}{e} \frac{(D^+ - D^-)\nabla c}{(D^+ + D^-)c}$$

This internal electric field derives from the junction potential

$$\Psi_{junc} = -\int_I^F \mathbf{E}_{int}.\mathbf{dx}$$

$$\Psi_{junc} = -\frac{RT}{\mathcal{F}} \frac{(D^+ - D^-)}{(D^+ + D^-)} \ln \frac{c_F}{c_i}$$

The diffusion flow can be rewritten in a more compact form

$$\mathbf{J}_i = -D_i \nabla c_i + \frac{kT}{e} \frac{\sum_j D_j Z_j \nabla c_j}{\sum_h Z_h^2 c_h D_h} Z_i e \frac{D_i}{kt} c_i$$

becomes then for a binary electrolyte

$$\mathbf{J}_+ = \mathbf{J}_- = -\frac{2 D^+ D^-}{D^+ + D^-} \nabla c$$

The coefficient of ∇c, $\mathcal{D}_{NH}$ is known as the Nernst-Hartley or mutual diffusion coefficient of the binary electrolyte

$$\mathcal{D}_{NH} = \frac{2 D^+ D^-}{D^+ + D^-}$$

4.3 Electrokinetic phenomena

We give in this section a very simplified presentation of electrokinetic phenomena, limiting us to the basic definitions and concepts [2].

At the end of this chapter, a bibliography is given.

Electrokinetic phenomena can be classified according to the nature of the acting field and to that of the response of the system. Another important classification comes from the geometry of the systems and from the nature and symmetry of the boundary conditions.

4.3.1 Electroosmosis

One first considers [3] the simple system consisting of an infinite charged plane, bearing fixed charges of one sign (they will be very often negative, as in silica, silicates and silicon, corresponding to glass, clay and chromatographic tubes). Obviously, mobile counterions of opposite sign will be located in the solution, in the vicinity of the fixed charges. They form together the double layer, which has been presented in the chapter dealing with electrostatics.

An external electric field $\mathbf{E}$ is applied parallel to the wall in the x direction, whereas the y direction, is counted positively perpendicular to the wall which is located at $y = 0$.

The basic equation of motion is the Navier Stokes Equation

$$\rho\frac{\partial \mathbf{v}}{\partial t} + \rho\mathbf{v}.\mathbf{grad}\,\mathbf{v} = \rho\mathbf{g} - \nabla P + \eta\Delta\mathbf{v} + \mathbf{f}^{el}$$

where $\mathbf{f}^{el}$ is the electric force density.

For low flow rates (low Reynolds numbers), and stationary flows the left-hand terms vanish.

In a first approach we neglect the gravity and pressure effects and get

$$\eta\Delta\mathbf{v} = -\mathbf{f}^{el} = -\rho^{el}\mathbf{E}$$

where ρ^{el} is the charge density and $\mathbf{E}$ the external applied electric field

$$\Delta\mathbf{v} = -\frac{1}{\eta}\rho^{el}\mathbf{E}$$

$$\mathbf{E} = E_x\mathbf{i}$$

$$\mathbf{v} = v_x\mathbf{i}$$

The external electric field is applied along the x direction, as well as the resulting velocity of the solvent.

$$\rho^{el} = \rho^{el}(y)$$

$$\eta\frac{\partial^2 v_x}{\partial y^2} = -\rho^{el}(y)E_x$$

Poisson's equation relates the charge density and the potential

$$\Delta \Psi = -\frac{\rho^{el}}{\varepsilon_0 \varepsilon_r}$$

$$\eta \frac{\partial^2 v_x}{\partial y^2} = \varepsilon_0 \varepsilon_r \frac{\partial^2 \Psi}{\partial y^2} E_x$$

This equation can be integrated as

$$v_x(y) = \frac{\varepsilon_0 \varepsilon_r}{\eta} \Psi(y) E_x + Ay + B$$

for symmetry reasons $A \equiv 0$.
The boundary conditions are:

$$\left. \begin{array}{ccc} y & = & 0 \\ v & = & 0 \\ \Psi & = & \Psi_s \end{array} \right\} \ at \ the \ interface$$

$$v_x(y = 0) = 0 = \frac{\varepsilon_0 \varepsilon_r}{\eta} \Psi_s E_x + B$$

then

$$B = -\frac{\varepsilon_0 \varepsilon_r}{\eta} \Psi_s E_x$$

$$v_x(y = \infty) = \frac{\varepsilon_0 \varepsilon_r}{\eta} E_x (\Psi_0 - \Psi_s)$$

where $\Psi(y = \infty) = \Psi_0$ is the potential of the solution.
Far from the plane one has

$$v_x(y = \infty) = v_{max}$$

$$v_{max} = \frac{\varepsilon_0 \varepsilon_r}{\eta} (\Psi_0 - \Psi_s) E_x$$

$$\Psi_0 - \Psi_s = \zeta$$

This defines the zeta potential as the potential difference in the double layer; we have then the simple relation

$$v_{max} = \frac{\varepsilon_0 \varepsilon_r}{\eta} \zeta E_x$$

for the electroosmotic velocity of the solvent.

The corresponding phenomenon (displacement of the solvent along a charged interface), is known as electroosmosis.

Charge density on the interface

In this paragraph the ζ potential will be presented by analogy to the charge of a plane capacitor, corresponding to the charged interface plus its counterions.

If we consider the wall as a capacitor C

$$C = \frac{Q}{U} = \frac{Q}{\zeta}$$

where Q is the charge of the interface and U the potential difference in the double layer

If the area of the double layer is A and its thickness δ

$$C = \varepsilon_0 \varepsilon_r \frac{A}{\delta}$$

we get

$$\zeta = \frac{Q\delta}{\varepsilon_0 \varepsilon_r A}$$

and

$$v_{max} = \frac{\sigma \delta}{\eta} E_x$$

where $\sigma = Q/A$ is the charge density on the interface.

For a dilute electrolyte, δ is close to $1/\kappa$, where $1/\kappa$ is the Debye length. This Debye length behaves as $1/\sqrt{C}$, C being the molar concentration of the added electrolyte.

The electroosmotic velocity has then a variation as $1/\sqrt{C}$ which is the basic feature of electroosmosis, in the total contradiction with the simple chemical argument where the flow of solvent would be due simply to the solvation of the mobile ions.

Electrophoresis

Electrokinetic phenomena occur not only for fixed interfaces and mobile counterions but also for mobile particles in suspension, either bearing a defined electric charge (and characterized by an electrophoretic mobility) or surrounded by a double layer, whatever could be the physical origin of this double layer. In this latter case the particle will undergo, under the action of an electric field, the same type of motion as in electroosmosis, with the noticeable difference that the motion is that of the particle in the fixed solvent.

The purpose of the present paragraph is to establish the basic relations between the electrophoretic and the electroosmotic mobility, for a spherical particle.

In electrophoresis, the velocity taken in a medium of viscosity η by a spherical particle of radius R and electric charge ze, under the action of an electric field $\mathbf{E}$ electric field is given by

$$\mathbf{v} = \omega\mathbf{F} = \frac{1}{6\pi\eta R}\mathbf{F} = \frac{ze\mathbf{E}}{6\pi\eta R}$$

Notice that the 6π coefficient is relative to sticking boundary conditions for the motion of the sphere in the solvent. This corresponds to wetting boundary conditions for the particle (good solvent).

Other features are possible for the contact between the moving sphere and the solvent. For the unwetting case, or slipping boundary conditions, the previous adimensional factor is replaced by 4π. This lower factor correspond to the lower friction in the slipping case. As simple way to express the relation to electroosmosis is as follows:

Assuming that the particle is equivalent to a spherical capacitor

$$C = \frac{Q}{U} = \frac{ze}{\zeta} = 4\pi\varepsilon_0\varepsilon_r R$$

The corresponding charge is

$$ze = 4\pi\zeta\varepsilon_0\varepsilon_r R$$

and the electroosmotic velocity for sticking conditions

$$\mathbf{v} = \frac{2}{3}\frac{\varepsilon_0\varepsilon_r\zeta\mathbf{E}}{\eta}$$

The adimensional factor $2/3$ being replaced by 1 for slipping conditions.

In this last case the result for the velocity is exactly that of simple electroosmosis for plane symmetry

$$\mathbf{v} = \frac{\varepsilon_0\varepsilon_r\zeta\mathbf{E}}{\eta}$$

4.3.2   Influence of boundary conditions

The measurement of ζ potential is in practice related to the particular set of boundary conditions used.

We distinguish the following cases

- open
- stationary
- closed

Open system

The liquid volume transported by time unit for a cylinder of diameter $\phi = 2r$ and length l is

$$\mathcal{D} = \frac{V}{t} = v\pi r^2$$

where v is the electrosmotic velocity
Substituting the expression for the electroosmotic velocity

$$\mathcal{D} = \frac{\varepsilon_0\varepsilon_r}{\eta}\zeta\pi r^2 \mathbf{E}$$

In a porous medium the determination of the electric field is not easy and $\mathbf{E}$ can be expressed in a different way. Taking the current intensity one obtains

$$I = \frac{U}{R} = \frac{El}{R}$$

since the measurable quantities are the intensity and the electric resistance and the conductance γ of the solution. Ohm's law writes simply

$$\mathbf{j} = \gamma\mathbf{E} = \frac{I}{S} = \frac{I}{\pi r^2}$$

and

$$I = \gamma\pi r^2 \mathbf{E}$$

and for the flow

$$\mathcal{D} = \frac{\varepsilon_0\varepsilon_r}{\eta}\zeta\frac{I}{\gamma}$$

This expression allows for the determination of ζ neither knowing the radius of the pores nor that of the capillaries. The measurement of a flow combined to that of intensity and electric conductance, allows for the determination of ζ potential in the porous medium with a charged double layer.

Open system: steady state

The experimental device is here very analogous to an osmosis apparatus, consisting in two open compartments communicating by a porous material. In the present case, an electric field is applied to the porous material and this substance is permeable to both solute and solvent particles. Moreover the porous material is supposed to be charged on surface. When a current passes through the porous medium, a difference Δh in liquid height takes place between the two compartments. This phenomenon ceases when the current stops. A difference Δp in pressure between the two compartments corresponds to Δh

$$\Delta p = \Delta h \rho g$$

Δp can be related to the flow $\mathcal{D}_P$ by Poiseuille's law

$$\mathcal{D}_P = \frac{V}{t} = \frac{\Delta p \pi r^4}{8 \eta l}$$

Since the system is open, by applying an electric field, the liquid goes from one side to the other, until the undergoing of a stationary state.

This steady state corresponds to a balance between the electroosmotic flow and the Poiseuille flow.

The electroosmotic flow is

$$\mathcal{D}_E = \frac{\varepsilon_0 \varepsilon_r}{\eta} \zeta \pi r^2 \mathbf{E}$$

then equating $\mathcal{D}_P$ and $\mathcal{D}_E$ yields

$$\Delta \mathbf{p} = 8 \frac{\varepsilon_0 \varepsilon_r \zeta l \mathbf{E}}{r^2}$$

The electric field can be replaced as for strictly open conditions in terms of the electric current, giving

$$\Delta_P = 8 \frac{\varepsilon_0 \varepsilon_r \zeta l I}{\gamma \pi r^4}$$

This method is very convenient to measure ζ potential for porous media, under laboratory conditions, whereas the strictly open conditions correspond more to *in situ* situations

Closed system: steady state

1. **Rectangular symmetry**

For closed systems, the electroosmotic flow will undergo the influence of the boundary conditions. Since on the wall the flow will be oriented by the direction of the field and the charge of the mobile counterions, on the bottom of the cell this flow will be redirected in the opposite direction. A counterflow will then be produced in the middle of the cell and solvent flow loops will be established between the electrodes.

The distance l between the two electrodes (direction of the field $\mathbf{E}$, along the horizontal x axis),will be supposed to be very large by respect to the other dimensions of the electrophoretic cell, respectively $2b$ in the vertical y direction and $2h$ in the other horizontal direction z, perpendicular to x.

$$2h << 2b << l$$

The origin of the coordinates is taken in the middle of the cell. The boundary effects such as the electrode reactions and the closing of the hydrodynamic loops are negligible in this region.

The velocity profile can be expressed as follows:

- in the vicinity of the wall in a boundary layer of thickness $\delta \simeq 1/\kappa$ the velocity of the solvent goes rapidly from zero to the Smoluchowski value

$$v_0 = \frac{\varepsilon_0 \varepsilon_r}{\eta} \zeta \, E_x$$

- after this rapid variation the velocity decreases in the direction of the middle of the cell, taking a negative value (by respect to that on the wall), in the middle. An important location will be that for which this velocity vanishes, somewhere between the wall and the middle of the cell. One of the goals of the present study will be to locate this location in $z(v = 0)$. This generates a plane for which the electroosmotic velocity is zero. This plane is known as the Smoluchowski plane.

- from the center of the cell to the opposite wall the phenomenon is exactly symmetrical: the velocity goes from a negative value to v_0 (chosen as positive), and then the velocity decreases to zero in the opposite double layer.

The velocity profile is then an even function by respect to the middle of the cell in the z direction.

The linearized steady state Navier-Stokes equation writes then as

$$\rho \mathbf{g} - \nabla P + \eta \Delta \mathbf{v} + \mathbf{f}^{el} = 0$$

The velocity is directed as the field, along the x axis and has non vanishing components only in this direction, except at the electrode boundaries. Moreover, since the y dimension of the cell is much more larger than that in z, the shorter hydrodynamic loops will be in this z direction.

This vector equation has the following components:

- x

$$\eta \left(\frac{\partial^2 v_x}{\partial y^2} + \frac{\partial^2 v_x}{\partial z^2} \right) - \frac{\partial P}{\partial x} = 0$$

- y

$$-\rho g - \frac{\partial P}{\partial y} = 0$$

- z

$$-\frac{\partial P}{\partial z} = 0$$

This last equation means simply that P is independent from z

The y component is simply

$$P = -\rho g y + cste(x)$$

Since the velocity v_x varies mainly as a function of z, the x component reduces then to:

$$\eta \frac{\partial^2 v_x}{\partial z^2} - \frac{\partial P}{\partial x} = 0$$

The solution of the problem will be obtained only by setting a supplementary hypothesis for the pressure gradient $\frac{\partial P}{\partial x}$.

The simplest hypothesis would be $\frac{\partial P}{\partial x} = 0$.

However this hypothesis would lead to

$$v_x(z) = Az + B$$

a profile in contradiction with the even character of the velocity.

The second simplest hypothesis for the pressure gradient is to consider it as a constant

$$\frac{\partial P}{\partial x} = p$$

and

$$\eta \frac{\partial^2 v_x}{\partial z^2} = p$$

The solution of this equation is

$$v_x(z) = \frac{p}{2\eta} z^2 + C_1 z + C_2$$

C_1 vanishes for symmetry reasons and C_2 will be determined by taking the condition

$$v_x(z = \delta) \simeq v_x(z = 0) = v_0$$

the electroosmotic velocity.

$$C_2 = v_0 - \frac{ph^2}{2\eta}$$

and

$$v_x(z) = \frac{p}{2\eta}(z^2 - h^2) + v_0$$

It remains to determine the value of the constant p.

The condition of zero total flow across the cell in the z direction writes

$$\int_h^h v_x(z)dz = 0$$

Substituting $v_x(z)$ gives

$$p = \frac{3v_0\eta}{h^2}$$

and

$$v_x(z) = \frac{v_0}{2h^2}(3z^2 - h^2)$$

This parabolic Poiseuille-like profile gives a stationary plane for

$$z_s = \pm\frac{h}{\sqrt{3}} = \pm0.577$$

When b is not infinitely larger than h the physics remains identical, but the stationary planes are called Komagata planes, and are located closer from the wall.

For $b/h = 10$ then $z_s/h = 0,612$ and for $b/h = 100$ then $z_s/h = 0,58$.

2. **Cylindrical system**

   The case of cylindrical symmetry remains physically identical to the previous one, but gives other mathematical difficulties.

   Taking as before the approximation of an infinite tube, and integrating the corresponding Navier-Stockes equation gives the velocity profiles

   $$v_x(z) = v_0(\frac{2r^2}{h_2} - 1)$$

   The Komagata surfaces are then located at

   $$z_s = \frac{h}{\sqrt{2}} = 0.71$$

4.3.3   Streaming potential

This phenomenon is the reverse effect from electroosmosis: a velocity field is applied along an interface and one observes the appearance of an electric field (or a an electric potential difference)associated to the flow.

The mathematical formulation of this effect will depend on the boundary conditions of the problem, as for electroosmosis [4].

In this paragraph only one example of such phenomena will be considered, in relation with the previous parts on electroosmosis.

Consider a cylindrical capillary with charged walls, through which a liquid is pushed under the action of an applied pressure Δp. A parabolic Poiseuille flow is then produced:

$$\mathbf{v}(r) = -\frac{R^2}{4\eta}(1 - (\frac{r}{R})^2)\nabla p$$

where R is the radius of the capillary of length l ($l \gg a$).

This flow drags counterions by respect to the fixed charged sites and then a streaming electric current is produced.

$$I_{str} = 2\pi \int_0^R v(r)\rho^{el}(r)dr = -\frac{\pi\nabla p}{2\eta}\int_0^R (R^2 - r^2)\rho^{el}(r)rdr$$

The exact solution of this problem with the correct symmetry will require the use of the cylindrical form of the Poisson equation for the charge density $\rho^{el}(r)$.

If the capillary is large enough (large κR), the one dimensional form of the Poisson equation can be used

$$\rho^{el}(r) = -\varepsilon_0\varepsilon_r\frac{d^2\Psi}{dy^2}$$

where y is the distance from the internal surface of the capillary.

Moreover, a very thin layer close to the interface contributes to the charge transport, since the bulk has no space charge. Then only the distance close to $r = R$ have to be considered:

$$R^2 - r^2 \simeq 2R(R - r) = 2Ry$$

Then the intensity can be rewritten as

$$I_{str} = \frac{\varepsilon_0\varepsilon_r\pi R}{\eta}\nabla p \int_0^R y\frac{d^2\Psi}{dy^2}Rdy = \frac{\varepsilon_0\varepsilon_r\pi R^2}{\eta}\nabla p \int_0^R y\frac{d^2\Psi}{dy^2}dy$$

This integral can be calculated by parts

$$\int_0^R y\frac{d^2\Psi}{dy^2}dy = \int_0^R yd\frac{d\Psi}{dy} = y\frac{d\Psi}{dy}|_0^R - \int_0^R \frac{d\Psi}{dy}dy$$

In the last equality the first term is zero because for $r = 0$, $\frac{d\Psi}{dy} = 0$ and for $r = R$, $y = 0$.

The last term is simply the ζ potential

$$\zeta = \Psi(r = 0) - \Psi(r = R)$$

The streaming intensity is then

$$\mathbf{I}_{str} = \frac{\varepsilon_0 \varepsilon_r \pi R^2}{\eta} \nabla p \zeta$$

This streaming intensity corresponds to an electric field $\mathbf{E}_{str}$ deriving from a streaming potential which can be expressed by considering the conductance γ of the capillary as for electroosmosis.

$$\mathbf{j}_{str} = \frac{\mathbf{I}_{str}}{S} = \gamma \mathbf{E}_{str}$$

$$\mathbf{E}_{str} = \frac{\varepsilon_0 \varepsilon_r}{\eta \gamma} \nabla p \zeta$$

Streaming potentials appear also in porous media and are now of fundamental importance for the study of geological materials and are used commonly to characterize their evolution such as the occurence of earth quakes and volcanic eruptions.

4.3.4   Applications of electrokinetic phenomena

Electrokinetic phenomena are sometimes hindered in practical applications such as filtration processes or separation by ion exchanging membranes [5] [6].

They are used as powerful separation methods in new methods such as capillary electrophoresis and electrokinetic decontamination. Those two applications will be rapidly presented, letting the interested reader explore these new tracks by himself.

Capillary electrophoresis

Capillary electrophoresis apparatus consist in a very long capillary tube, generally in silica or in a siliconated material, whose diameter is of some μm. A very strong electric field is applied to the extremities of this device which contains an electrolyte solution, generally dilute.

The strong electric field acts on the mobile positive counterions located on the wall of the negatively charged capillary tube, creating a huge electroosmotic solvent flow.

If one introduces by a chromatographic injection technique, a solution of particles, they will be submitted to the electric field and take a velocity which is the balance between their own electric mobility and the electroosmotic flow. If two particles i and j have intrinsic mobilities $\omega_i = \omega - \Delta\omega$ and $\omega_j = \omega + \Delta\omega$, under the application of the electric field they can move in opposite directions if

$$\omega = -\frac{\varepsilon_0 \varepsilon_r \zeta}{\eta}$$

i.e. if the intrinsic mobility of the particles is close to the opposite of the electroosmotic velocity.

In the present example one will separate negatively charged particles. In order to apply this technique to positive particles, it should be noticed that the charge of the wall can be inverted by the following technique. By putting in the solution an ionic surfactant positively charged (e.g. dodecyl trimethylammonium bromide), the polar heads of the surfactant will neutralize the negative sites of the capillary wall, forming a surfactant layer on it. Other surfactant molecules will be trapped by the hydrophobic tails, constructing a bilayer on it. The net charge density of the wall will then be positive, creating an electroosmotic flow directed in a direction opposite to the previous one, which can be used to separate cations.

This constitutes the schematic principle of operation for capillary electrophoresis, which is now a commonly used chromatographic technique.

Electrokinetic decontamination

A soil like clay contaminated by heavy metal pollution, can be considered as a porous medium containing water and ions adsorbed (heavy metals cations) on negatively charged sites (silicates). By applying a moderate electric field and injecting an appropriate solvent, one will obtain an electroosmotic flow of solvent containing the heavy metal ions to eliminate.

This method has been applied with some success, to the removal of As, Cd, Hg, Cs, Pu from contaminated industrial sites. since the required fields are relatively low (some volts by meter), the main expanses will come from the cost of the solvent.

Another difficulty in old polluted industrial sites comes from the fact that hidden solid metallic objects and residues are often present in the soil.

A last application of electrokinetic phenomena is the concentration of industrial muds. Here also, with a small expanse in energy, elimination of water by electroosmosis allows to concentrate muds from 70% of water (a liquid like mud), to 20% of water (a solid, clay rock like).

The role of electrokinetic phenomena in environment and environmental technologies is then crucial and promising.

Bibliography

[1] P. Turq, J. Barthel and M. Chemla *Transport, Relaxation and Kinetic Processes in Electrolytes* Lecture Notes in Chemistry, 57, Springer Verlag, Berlin, 1992.

[2] J. Lyklema *Fundamentals of Interface and Colloid Science*, Academic Press, London, 1995.

[3] M. von Smoluchowski *Bull. Int. Acad. Sci. Cracovie* (1903) 184.

[4] M. von Smoluchowski *Z. Phys. Chem.* 92 (1918) 129.

[5] A. Lehmani, P. Turq, M. Perie, J. Perie and J.P. Simonin *J. Electroanal. Chem.* 428 (1997) 81.

[6] A. Lehmani, O. Bernard and P. Turq *J. Stat. Phys.* 89 (1997) 379.

Chapter 5

Description of electrolyte transport using the MSA for simple electrolytes, polyelectrolytes and micelles

Description of the concentration dependence of the transport coefficients of electrolytes is one of the oldest open problem in physical chemistry. Since the early papers of Onsager *et al* in 1926 [1] and in 1932 [2], where limiting laws for the conductance were given, and later extended to the self diffusion for single ions [3] and ionic mixtures [4], progress has been difficult.

In 1957 Onsager *et al.* [4] made an attempt to extend the validity of the conductivity limiting law to higher concentrations, using the Debye-Hückel equilibrium pair distribution functions [6]. At the same level, concentration dependence for self-diffusion was obtained [7, 8].

Ebeling *et al* [9, 10] used the restricted primitive model (equal size ions in a dielectric continuum) to describe the variation of conductance with concentration. They used MSA distribution functions to compute the relaxation contribution.

Some groups have recently formulated a linear response theory in which Onsager's continuity equations were combined with the MSA equilibrium correlation functions, using the Green's response functions formalism. They have used a primitive model description, in which solvent effects are averaged. This yields concentration independent potentials, generally valid in the 0-1 M concentration range. This approach has been applied to self-diffusion [11], acoustophoresis [12], conductance of two simple ionic species [13] for non-associating electrolytes. The treatment has been extended to associated electrolytes for conductance and self-diffusion [14, 15], using a chemical model of association. They also extended the theory to the conductance of electrolytes with three ionic species and to ionic micellar systems.

We present here the basis of the development and the main results.

5.1   General theory

The dominant forces that determine deviations from ideal behaviour of transport processes in electrolytes are the relaxation and electrophoretic forces [16]. The first of these forces was discussed by Debye [6, 17]. When the equilibrium ionic distribution is perturbed by some external force in an ionic solution, electrostatic forces appear, which will tend to restore the equilibrium distribution of the ions. There is also a hydrodynamic effect. It was first discussed by Onsager [2, 3]. Different ions in a solution will respond differently to external forces, and will thus tend to have different drift velocities: The hydrodynamic (friction) forces, mediated by the solvent, will tend to equalize these velocities. The electrophoretic (hydrodynamic) correction can be evaluated by means of Navier-Stokes equation [18, 19]. Calculating the relaxation effect requires the evaluation of the electrostatic drag of the ions by their surroundings. The time lag of this effect is known as the *Debye relaxation time*.

Before going to the calculations, the various forces, namely the relaxation and electrophoretic forces, which are corrections to the main driving forces, are schematically depicted in fig. 5.1, when an electric field is applied.

We start with the hydrodynamic continuity equations [16]

$$-\frac{\partial f_{ij}}{\partial t} = \nabla_1(f_{ij}v_{ij}) + \nabla_2(f_{ji}v_{ji}) \tag{5.1}$$

where v_{ij} is the velocity of an ion j in the vicinity of an ion i and f_{ij} is the two–particle density, related to the pair distribution function $g_{ij}(r, t)$

$$f_{ij}(r, t) = \rho_i\rho_j g_{ij}(r, t) \tag{5.2}$$

Where r is the distance between ions i and j, t is the time and ρ_i is the particle density (ions/volume).

The pair distribution function is related to the total distribution function $h_{ij}(r, t)$

$$g_{ij}(r, t) = 1 + h_{ij}(r, t) \tag{5.3}$$

In the linear response theory, the total pair distribution is expressed as the sum of an equilibrium part (superscript [0]) and a part that is proportional to the external perturbation (superscript [1]).

$$h_{ij}(r, t) = h_{ij}^o(r) + h_{ij}^1(r, t) \tag{5.4}$$

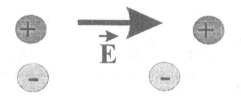

Applied External Field

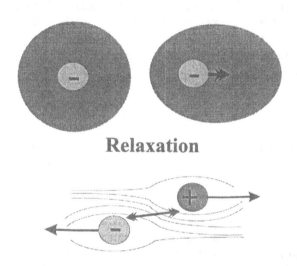

Relaxation

Hydrodynamic Interaction

Figure 5.1: Schematical representation of Relaxation and Electrophoretic forces. The relaxation force is an electrical force which appears when the local electric equilibrium is disturbed. In this case, the electrical perturbation is due to the ionic atmosphere deformation. The electrophoretic force is caused by hydrodynamic interactions which tend to equalize the velocities of the two ions.

The velocity v_{ji} of an ion of species i in the vicinity of an ion of species j is given by

$$v_{ji} = v_i^s + \omega_i (\mathbf{K}_{ji} - k_B T \nabla \ln f_{ji}) \tag{5.5}$$

where the generalized mobility ω_i of the ion i is related to its diffusion coefficient D_i by the relation $\omega_i = D_i / k_B T$ (k_B is the Boltzmann constant and T is the absolute temperature), v_i^s is the average relative velocity of the solvent with respect to the ion of species i and $\mathbf{K}_{ji}$ is the force acting on an ion of species i in the neighbourhood of species j

$$\mathbf{K}_{ji} = \mathbf{k}_i \left(1 + \frac{\delta k_i}{k_i}\right) - e_i \nabla \psi_j \tag{5.6}$$

In equation (5.6) $\mathbf{k}_i$ is the external (diffusive or electrostatic) force on an ion i. In the conductance case, the external force is the applied electrostatic field. For the self–diffusion of a tracer, the external force is the gradient of the chemical potential created by the gradient of (isotopic) concentration of the diffusing tracer. In the acoustophoresis, it is the sound pressure wich induces a differential displacement of each ion so that an induced electric field appears.
δk_i is the relaxation force [20, 21]

$$\delta k_i = -\sum_j \rho_j \int_0^\infty \nabla (V_{ij}^{Cb} + V_{ij}^{hs}) h_{ij}^1 d\mathbf{r} \tag{5.7}$$

where V_{ij}^{Cb} is the Coulomb potential and V_{ij}^{hs} is the hard sphere potential

$$V_{ij}^{hs} = \begin{cases} \infty & \text{if } r < \sigma_{ij} \\ 0 & \text{if } r \geq \sigma_{ij} \end{cases} \tag{5.8}$$

σ_{ij} is the sum of the effective radii of the two ions i and j, $\sigma_{ij} = (\sigma_i + \sigma_j)/2$. Let us denote by ψ_j the electrostatic potential around an ion j. In the linear response theory it can be expressed as the sum of an equilibrium part and a part that is proportional to the external perturbation.

$$\psi_j(r, t) = \psi_j^0(r) + \psi_j^1(r, t) \tag{5.9}$$

In the absence of external forces the pair distribution functions (pdf) are the equilibrium ones; They must satisfy the symmetry relation

$$g_{ij}^0 = g_{ji}^0 \tag{5.10}$$

and the excluded volume condition

$$g_{ij}^0(r) = 0 \qquad \text{if} \qquad r \leq \sigma_{ij} \tag{5.11}$$

The equilibrium pdf's can be computed nowadays very accurately by theories such as HNC [22] and some of its improved versions [23]. The MSA [24, 25] is the simplest theory that satisfies all of the above conditions. It is the Debye-Hückel theory, but solved with the condition (5.11) for all pairs of ions. The final result introduces a new screening parameter Γ (intead of the Debye screening parameter κ) which is calculated from an algebraic equation [26, 27, 28]. It was found that the MSA is sufficiently accurate, in all of the studied cases.

The basic assumption is that the excess non-equilibrium potential ψ_j^1 also satisfies the Poisson equation even if the equilibrium state is not established. It means that:

$$\Delta\psi_j^1 = -\frac{1}{\epsilon_0\epsilon_r}q_j^1(r) = -\frac{1}{\epsilon_0\epsilon_r}\sum_i \rho_i e_i h_{ji}^1 \tag{5.12}$$

where $q_j^1(r)$ is the non equilibrium charge excess density around an ion j. In this equation the permittivity of the vacuum ϵ_0 is $8.8542 \cdot 10^{-12}$ F m^{-1} in SI units, ϵ_r is the relative dielectric permitivity, and e_j the charge of the ion j.

The continuity equations are expressed as the sum of equilibrium terms plus a perturbation. The form of this perturbation depends on the type of transport phenomenon studied. In all cases the continuity equations can be written in the form of an inhomogeneous differential equation such as:

$$\left(\Delta - \kappa_r^2\right) h(r) = F(\mathbf{r}, \mathbf{k}) \tag{5.13}$$

where κ_r is a dynamic screening parameter that depends on the individual ionic mobilities and $F(\mathbf{r}, \mathbf{k})$ is a generalized driving force.

For self-diffusion and conductivity of binary ionic mixtures, it was found that mean electrostatic and hard sphere diameters [28, 29] reduce the complexity of the formulas, without reducing the accuracy of the equations. The mean electrostatic diameter is given by:

$$\sigma = \left(\sum_i z_i^2\rho_i\sigma_i\right) / \left(\sum_i z_i^2\rho_i\right) \tag{5.14}$$

Equation (5.13) can be solved up to first–order, or second–order in the perturbing force. Two kinds of terms are obtained:

i) terms yielding the limiting laws in $\sqrt{c}$ when the Debye Hückel equilibrium pair distribution functions are used [6].

ii) higher order terms in concentration with the same distribution functions (second–order terms).

The contribution of the second-order terms is always small for concentrations lower than 0.5 M for most 1-1 electrolytes in water. It can then be neglected as a first approximation.

We discuss particular transport processes in the next sections.

5.2 Self-diffusion

In this case the relaxation forces are the only relevant contributions and the hydrodynamic interactions do not play a significant role in the range of concentration of 0-2 M except for low viscosity solvents [30]. Neglecting hydrodynamic interactions, equation (5.13) gives, to the first order approximation:

$$\Delta X_1^1 - \kappa_{d_1}^2 X_1^1 = \frac{k_1 D_1^o \cos \Theta}{k_B T} \sum_{j=2}^{3} \frac{\rho_j e_j}{D_1^o + D_j^o} \frac{d}{dr} g_{j1}^o(r) \tag{5.15}$$

where Θ is the angle between de gradient concentration and the position vector and

$$X_1^1 = \sum_j \rho_j e_j h_{j1}^1 \tag{5.16}$$

Here the subscript 1 indicates the tracer ion. The function δk_1^1 is related to X_1^1 by the first-order non-equilibrium correlation functions.

The solution of equation (5.15) yields the following expression

$$\frac{\delta k_1^1}{k_1} = \frac{-Z_1^2 e^2 \left(\kappa^2 - \kappa_{d_1}^2 \right)}{6 \epsilon k_B T \sigma \left(1 + \Gamma \sigma \right)^2} \frac{\left(1 - e^{-2\kappa_{d_1} \sigma} \right)}{\left[\kappa_{d_1}^2 + 2 \Gamma \kappa_{d_1} + 2 \Gamma^2 \left(1 - e^{-\kappa_{d_1} \sigma} \right) \right]} \tag{5.17}$$

where the MSA Laplace transform of h is used [27] and $1/\Gamma$ is the MSA screening length parameter [26, 27]. It is convenient to use the average diameter [28] σ to compute this parameter from Debye's κ:

$$\kappa^2 = \frac{4\pi e^2}{\epsilon k_B T} \sum_n \rho_n Z_n^2 = 4\Gamma^2 \left(1 + \Gamma \sigma \right)^2 \tag{5.18}$$

Furthermore, the dynamic Debye parameter is

$$\kappa_{d_1}^2 = \frac{4\pi e^2}{\epsilon k_B T} \sum_n \frac{\rho_n Z_n^2 D_n^o}{D_1^o + D_n^o} \tag{5.19}$$

The formal expression of the diffusion coefficient becomes then

$$D_1 = D_1^o \left(1 + \frac{\delta k_1^1}{k_1^1} \right) \tag{5.20}$$

5.3 conductance of two simple ionic species

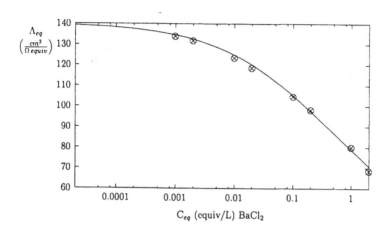

Figure 5.2: Equivalent conductance of aqueous $BaCl_2$ solutions at 25°C as a function of the molarity. $\otimes$: experimental values taken from ref [37, 38]. (—): theoretical values predicted by the MSA expression with average diameter $\sigma = 3.55$ Å ($\sigma_{cr} = 3.15$ Å).

In a conducting solution, anions and cations move in opposite directions, and both the relaxation effects and the hydrodynamic interactions must be taken into account.

5.3.1 Relaxation

In this case equation (5.13) becomes

$$\Delta h_{ji}^1 - \kappa_q^2 h_{ji}^1 = \frac{e_i D_i^o - e_j D_j^o}{k_B T (D_i^o + D_j^o)} \mathbf{E} \nabla g_{ji}^o \tag{5.21}$$

The solution of equation (5.21) yields the following expression

$$\frac{\delta k_1^1}{k_1} = \frac{\delta k_2^1}{k_2} = \frac{\delta E}{E} = \frac{-\kappa_q^2 e^2 \mid Z_1 Z_2 \mid}{3\epsilon k_B T \sigma \left(1 + \Gamma \sigma\right)^2} \left[\sinh(\kappa_q \sigma) - \frac{\epsilon k_B T}{e_i e_j} \left(\cosh(\kappa_q \sigma) - \frac{\sinh(\kappa_q \sigma)}{\kappa_q \sigma} \right) \right]$$

$$\times \frac{e^{-\kappa_q \sigma}}{\left[\kappa_q^2 + 2\Gamma \kappa_q + 2\Gamma^2 \left(1 - e^{-\kappa_q \sigma} \right) \right]} \tag{5.22}$$

with

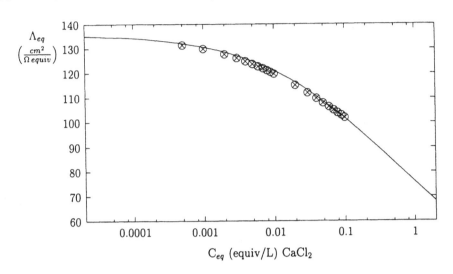

Figure 5.3: Equivalent conductance of aqueous $CaCl_2$ solutions at 25°C as a function of the molarity. $\otimes$: experimental values taken from ref [37, 38]. (—): theoretical values predicted by the MSA expression with average diameter $\sigma = 3.64$ Å ($\sigma_{cr} = 2.80$ Å).

$$\kappa_q^2 = \frac{4\pi e^2}{\varepsilon k_B T} \frac{\rho_i Z_i^2 D_i^o + \rho_j Z_j^2 D_j^o}{D_i^o + D_j^o} \tag{5.23}$$

and

$$\mathbf{k}_i = Z_i e \mathbf{E} \tag{5.24}$$

where Z_i is the valence of the ion j and e is the elementary charge ($\equiv$ $1.6\,10^{-19}\,C$).

5.3.2   Electrophoretic effect

The electrophoretic effect is due to hydrodynamic interactions between the ions (cf fig. 5.1). To the first order approximation one gets :

$$\delta \mathbf{v}_i^{el} = \sum_j \rho_j e_j \int_0^\infty h_{ij}^0(r) \mathcal{T} \mathbf{E}\, d\mathbf{r} \tag{5.25}$$

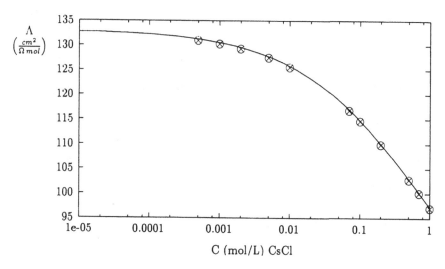

Figure 5.4:   conductance of aqueous CsCl solutions at 18°C as a function of the molarity. $\otimes$: experimental values taken from ref [36]. (—): theoretical values predicted by the MSA expression with average diameter $\sigma = 3.48$ Å ($\sigma_{cr} = 3.48$ Å).

where $\mathcal{T}$ is the Oseen tensor [31]

$$\mathcal{T}(r) = \frac{1}{8\pi\eta_o r}\left(\mathcal{I} + \frac{\mathbf{r}\otimes\mathbf{r}}{r^2}\right) \tag{5.26}$$

where η_o is the viscosity of pure solvent and $\mathcal{I}$ is the unity tensor.
In this case the MSA pair distribution functions yield a simple extension of Henry's law for electrophoretic mobility [18]

$$\frac{\delta u_i^{el}}{u_i^o} = \frac{-k_B T}{3\pi\eta_o D_i^o}\frac{\Gamma}{1+\Gamma\sigma} \tag{5.27}$$

Where u_i^o and D_i^o are the electrophoretic mobility and the diffusion coefficient of an ion i at infinite dilution.

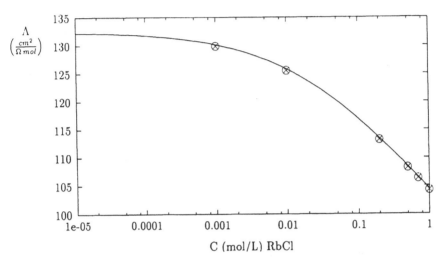

Figure 5.5:   conductance of aqueous RbCl solutions at 18°C as a function of the molarity. (—): theoretical values predicted by the MSA expression with average diameter $\sigma = 4.82$ Å ($\sigma_{cr} = 3.38$ Å).

5.3.3   Expression of the conductance and comparison with experiment

The total equivalent conductance is $\Lambda = \sum_i \lambda_i$ where

$$\lambda_i = \lambda_i^o \left(1 + \frac{\delta u_i^{el}}{u_i^o}\right)\left(1 + \frac{\delta E}{E}\right) \tag{5.28}$$

with λ_i^o the individual equivalent conductance of the ith ion at infinite dilution,

$$\lambda_i^0 = \frac{D_i^0 \mid z_i \mid F^2}{RT} \tag{5.29}$$

where F is the Faraday constant, z_i the charge number of ion i and R is the gas constant.

At this approximaton order (first-order term and mean cut-off distance) the conductance is then given by:

$$\Lambda_{eq} = \sum_{i=1}^{2} \lambda_i^0 \left(1 + \frac{\delta u_i^{el}}{u_i^o}\right) \left(1 - \frac{k_B T \kappa}{6\pi\eta \, D_i^0 \, (1 + \Gamma\sigma)^2}\right) \tag{5.30}$$

This quantity is experimentally known with high precision. The closest-approach distance σ is calculated by fitting to the expression for the equivalent conductivity the conductance data of a given salt. The fitted σ is, in all studied cases, slightly larger than the sum of the anion's and the cation's crystallographic radii σ_{cr}. At low concentrations ($< 10^{-5}$ mol/L), theoretical results converge asymptotically to the ideal behaviour.

Next figures give some examples for simple non-associated electrolytes. Concerning 2-1 salts, equivalent concentration and conductivity was chosen.

5.4 conductance in electrolyte mixtures: Case with three simple ionic species

Conductance of solutions containing more than two ion species was first theoretically studied in a systematic way by Onsager and Kim in 1957 [4].

They used Debye-Hückel equilibrium pair distribution functions whose range of validity is lower than 0.01 mol/L. Also, their theoretical expressions (limiting laws), are in agreement with experiment up to salt concentrations of 0.01 mol/L.

Later Quint and Viallard [32, 33] were able to extend this limit to 0.1M, by introducing finite ionic size corrections (extended limiting law). This change allowed them to increase by a factor of 10, the range of validity of the Onsager-Kim treatment with respect to concentration.

It is crucial to have good equilibrium pair distribution functions up to concentrations at least as high as 1M, in order to be able to calculate transport properties up to these concentrations. It is the case when the Laplace transforms of the MSA distribution functions [27] are used in place of the Debye-Hückel distribution functions.

In this section, new results [34] concerning the theoretical description of conductance for solutions containing three simple ionic species are presented. It is the application of a transport model to the conductance of ionic mixtures, with MSA equilibrium pair distribution functions. The model has been tested on NaCl/KCl and NaCl/MgCl$_2$ mixtures and reproduces well the experimental values within a wide concentration range.

5.4.1 Evaluation of the relaxation terms

The continuity equation to first order perturbation is

$$\Delta h_{ji}^1 + \frac{e_i \omega_i \Delta \psi_i^1 - e_j \omega_j \Delta \psi_j^1}{k_B T \left(\omega_i + \omega_j\right)} = \mathbf{A}_{ji} \nabla h_{ji}^0 \tag{5.31}$$

$$\mathbf{A}_{ji} = \frac{\mathbf{k}_i \omega_i - \mathbf{k}_j \omega_j}{k_B T \left(\omega_i + \omega_j\right)} \tag{5.32}$$

where $\mathbf{k}_i$ is the external electrostatic force acting on ion i.

We consider only two-body interactions; moreover, symmetry requires that: $h_{ji}^1 = -h_{ij}^1$ and $h_{ii}^1 = 0$ [2]. The problem is then reduced to a system of three equations with three unknowns. h_{12}^1, h_{13}^1 and h_{23}^1 or alternatively ψ_1^1, ψ_2^1 or ψ_3^1 since those last quantities are related by Poisson's equation(5.12).

Equations (5.12) and (5.31) lead to a system of three differential equations, one for each one of the three species j:

$$\Delta Y_j - \kappa^2 \bar{\omega} \left(Y_j \sum_{l=1}^{3} (t_l/b_{lj}) - t_j \sum_{l=1}^{3} (Y_l/b_{lj}) \right) = T_j \tag{5.33}$$

Where

$$Y_i = n_i e_i \Delta \psi_i^1 \tag{5.34}$$

$$t_i = n_i e_i^2 \omega_i / \left(\sum_{l=1}^{3} n_l e_l^2 \omega_l \right) \tag{5.35}$$

$$\kappa^2 = \frac{1}{k_B T \varepsilon_o \varepsilon_r} \sum_{l=1}^{3} n_l e_l^2 \tag{5.36}$$

$$b_{kj} = \dot{\omega}_k + \omega_k \tag{5.37}$$

$$T_j = \frac{1}{\varepsilon_o \varepsilon_r} \sum_{l=1}^{3} \left[\frac{a_l}{b_{lj}} \left(\mathbf{k}_l \omega_l - \mathbf{k}_j \omega_j \right) \nabla \left(a_j h_{jl}^o \right) \right] \tag{5.38}$$

$$a_i = n_i e_i \tag{5.39}$$

$$\bar{\omega} = \sum_{i=1}^{s} \mu_i \omega_i \tag{5.40}$$

$$\omega_i = D_i^o / (k_B T) \tag{5.41}$$

Taking into account

$$\sum_{j=1}^{3} Y_j = 0 \tag{5.42}$$

the problem reduces to a system of two equations which can be solved by Fourier transformations.

If we assume that the direction of the external applied field is along the axis z ($\mathbf{E} = E\,\mathbf{z}$) the projection of the forces along z is given by :

$$\left(\delta k_1^{rel}\right)_z = \frac{C_1 E}{3 n_1} \sum_{l=1}^{3} \left\{ L_{2l} f_{12} \left(G_{2l}\left(\kappa_+\right) - G_{2l}\left(\kappa_-\right)\right) \right.$$
$$\left. + L_{1l} \left(G_{1l}\left(\kappa_+\right)\left(f_{11} - \kappa_+^2\right) + G_{1l}\left(\kappa_-\right)\left(\kappa_-^2 - f_{11}\right)\right) \right\}$$

$$\left(\delta k_2^{rel}\right)_z = \frac{C_1 E}{3 n_2} \sum_{l=1}^{3} \left\{ L_{1l} f_{21} \left(G_{1l}\left(\kappa_+\right) - G_{1l}\left(\kappa_-\right)\right) \right.$$
$$\left. + L_{2l} \left(G_{2l}\left(\kappa_+\right)\left(f_{21} - \kappa_+^2\right) + G_{2l}\left(\kappa_-\right)\left(\kappa_-^2 - f_{21}\right)\right) \right\}$$

$$\left(\delta k_3^{rel}\right)_z = \frac{C_1 E}{3 n_3} \sum_{l=1}^{3} \left\{ L_{1l} \left(G_{1l}\left(\kappa_+\right)\left(\kappa_+^2 - (f_{11} + f_{21})\right) + G_{1l}\left(\kappa_-\right)\left((f_{11} + f_{21}) - \kappa_-^2\right)\right) \right.$$
$$\left. + L_{2l} \left(G_{2l}\left(\kappa_+\right)\left(\kappa_+^2 - (f_{12} + f_{22})\right) + G_{2l}\left(\kappa_-\right)\left((f_{12} + f_{22}) - \kappa_-^2\right)\right) \right\}$$

where

$$C_1 = 1/\left(\varepsilon_o \varepsilon_r k_B T \left(\kappa_-^2 - \kappa_+^2\right)\right) \tag{5.43}$$

κ_- and κ_+ are the roots of the determinant of the system eqs. (5.33) and (5.42). Furthermore

$$L_{jl} = \frac{e_l D_l - e_j D_j}{D_l + D_j} n_l e_l n_j e_j \tag{5.44}$$

and

$$f_{11} = \left(\kappa^2 \bar{\omega} / \left(\omega_2 + \omega_3\right)\right)\left(1 + t_1\left(\omega_3 - \omega_1\right)/\left(\omega_1 + \omega_2\right)\right) \tag{5.45}$$
$$f_{12} = -\kappa^2 \bar{\omega} t_1 \left((1/b_{13}) - (1/b_{12})\right) \tag{5.46}$$
$$f_{21} = -\kappa^2 \bar{\omega} t_2 \left((1/b_{23}) - (1/b_{12})\right) \tag{5.47}$$
$$f_{22} = \left(\kappa^2 \bar{\omega} / \left(\omega_1 + \omega_3\right)\right)\left(1 + t_2\left(\omega_3 - \omega_2\right)/\left(\omega_1 + \omega_2\right)\right) \tag{5.48}$$

and

$$G_{ij}\left(\kappa\right) = G_{ji}\left(\kappa\right) =$$
$$\frac{e_i e_j \sinh\left(\kappa \sigma_{ij}\right) \exp\left(\kappa \sigma_{ij}\right)}{\left\{\sigma_{ij} 4\pi\varepsilon_o\varepsilon_r k_B T \left(1 + \Gamma\sigma_i\right)\left(1 + \Gamma\sigma_j\right)\left[\kappa^2 + 2\Gamma\kappa + 2\Gamma^2\left(1 - \frac{1}{\alpha^2}\sum_{k=1}^{3} n_k a_k^2 \exp\left(-\kappa\sigma_k\right)\right)\right]\right\}}$$

For the very asymmetric case, individual ionic radii must be used to calculate Γ with the equations of Blum et al. [26, 27]

$$4\Gamma^2 = \frac{e^2}{\varepsilon_o \varepsilon_r k_B T} \sum_{i=1}^{3} n_i \left[\left(z_i - \frac{\pi}{2\Delta} P_n \sigma_i^2\right) / \left(1 + \Gamma\sigma_i\right)\right]^2 \tag{5.49}$$

where

$$P_n = (1/\Omega) \sum_{k=1}^{3} \left(n_k \sigma_k z_k / (1 + \Gamma \sigma_k) \right) \tag{5.50}$$

$$\Omega = 1 + (\pi/2\Delta) \sum_{k=1}^{3} \left(n_k \sigma_k^3 / (1 + \Gamma \sigma_k) \right) \tag{5.51}$$

$$\Delta = 1 - (\pi/6) \sum_{k=1}^{3} n_k \sigma_k^3 \tag{5.52}$$

5.4.2   Hydrodynamic correction

The hydrodynamic (or electrophoretic) contribution can be written as:

$$\delta \mathbf{k}_i^{hyd} = \zeta_i \, \delta \mathbf{v}_i^{hyd} \tag{5.53}$$

where the correction on the velocities is [13, 31]:

$$\delta \mathbf{v}_i^{hyd} = - \sum_{j=1}^{2} n_j \int_{\mathbf{r}} h_{ij}^0(\mathbf{r}) \, \mathcal{T}(\mathbf{r}) \, \mathbf{k}_j \, d\mathbf{r} \tag{5.54}$$

with $\mathcal{T}(\mathbf{r})$ the Oseen's tensor.

Bernard et al.[18] made the evaluation of this contribution from the excess internal energy [26]. We report their result: here, In this case, the sum is made over three species instead of two.

$$\delta \mathbf{v}_i^{hyd} = - \frac{e_i \mathbf{E}}{3\pi \eta_o} \left(\frac{\Gamma}{1 + \Gamma \sigma_i} + \frac{\pi}{2\Delta} \frac{P_n \sigma_i}{z_i (1 + \Gamma \sigma_i)} + \frac{\pi}{z_i} \sum_j n_j z_j \sigma_{ij}^2 \right) \tag{5.55}$$

In this expression, the last two terms in this expression arise from the asymetry of size of the ions.

5.4.3   Explicit expression for the conductance and comparison with experiment

The expression for the specific conductivity is

$$\chi_{sp} \left(\Omega^{-1} cm^{-1} \right) = \frac{10 e^2 N_A}{k_B T} \sum_{i=1}^{3} c_i D_i^o z_i^2 \left(1 + \delta v_i^{hyd} / v_i^o \right) \left(1 + \delta k_i^{rel} / k_i \right) \tag{5.56}$$

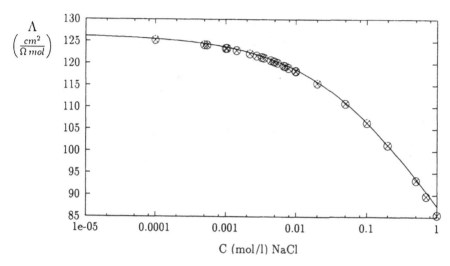

Figure 5.6:   Conductance of aqueous NaCl solutions at 25°C as a function of the salt molarity. ($\otimes$) : experimental values taken from ref [41, 42, 43, 44, 45]. (—): theoretical values predicted by the MSA expression with radii of $r_{Na^+} = 1.3$Å and $r_{Cl^-} = 1.81$Å.

where c_i is the molar concentration of ion i, D_i^o its diffusion coefficient at infinite dilution and v_i^o is the velocity at infinite dilution (without ionic strength correction),

$$v_i^o = e_i E D_i^o / (k_B T) \tag{5.57}$$

We express now the conductance as the specific conductivity χ_{sp} divided by the common ion concentration c_2 (molar conductivity).

$$\Lambda \left(cm^2 \Omega^{-1} mol^{-1}\right) = 1000 \chi_{sp} / c_2 \tag{5.58}$$

Some examples of the application of this theory are given in the next figures.

There are very few experimental data available in the literature for the conductivity of electrolyte mixtures at moderate and high concentration. This fact is certainly a consequence of lack of theoretical models in both the dilute and concentrated regions. We present here results for two mixtures, NaCl/KCl and NaCl/MgCl$_2$.

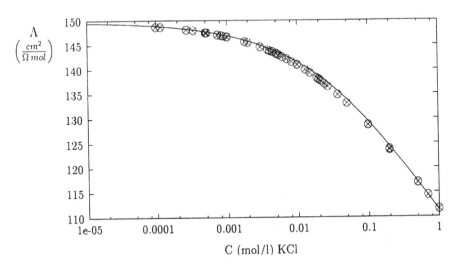

Figure 5.7: Conductance of aqueous KCl solutions at 25°C as a function of the salt molarity. ($\otimes$) : experimental values taken from ref [44, 46, 47, 48]. (—): theoretical values predicted by the MSA expression with radii of $r_{K^+} = 1.58$Å and $r_{Cl^-} = 1.81$Å.

In order to describe properly assymmetric systems, three distances are considered for each mixture. Two sets of values are presented: the crystallographic radii (minimum size values) and the radii r_i ($\sigma_i = 2r_i$) which give better reproduction of experimental results for single pairs. The three figures (fig. 5.6-5.8) are the results obtained for the conductivity of NaCl, KCl and MgCl$_2$ in water. Whereas the Onsager treatment [4] is limited to 0-0.01 mol/L concentration range, this theory is in agreement with experiment in the range 0-1 mol/L using the radii: $r_{Na^+} = 1.3$Å , $r_{Cl^-} = 1.81$Å and $r_{Mg^{2+}} = 1.82$Å . These values were then used to describe the conductivity of the two mixtures. Some improvementss in the high ionic strength domain could be obtained with a second order expansion.

Concerning the mixture NaCl/KCl, figures 5.9-5.11 present the conductance as a function of the proportion of one of the salts (KCl), for different total salt concentrations. One test of the theory is the asymptotic convergence of the results at low

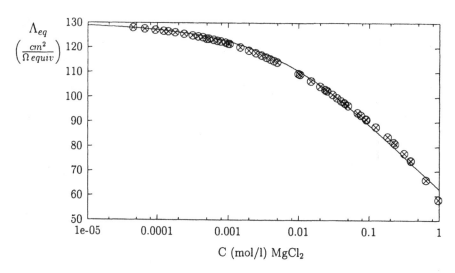

Figure 5.8: Equivalent conductance of aqueous $MgCl_2$ solutions at 25°C as a function of the salt molarity. $(\otimes)$: experimental values taken from ref [49, 50, 51]. $(-)$: theoretical values predicted by the MSA expression with radii of $r_{Mg^{2+}} = 1.82$Å and $r_{Cl^-} = 1.81$Å.

concentration to those of Onsager-Kim. Both results must verify the ideal limiting law at very low concentration. We present therefore Onsager-Kim limiting law for the lowest concentrations, as well as the ideal law (no interaction) (fig. 5.9). All curves merge at vanishing concentration.

It can be noticed that on figure 5.9 both choices of distance are in good agreement with experiment, due to the small contribution of hard sphere interactions at low concentrations. For a total concentration of 0.5 molal (fig. 5.10) and 0.75 molal (fig. 5.11), we observe a notable difference between the two sets of ionic radii, whereas the ideal and limiting law models are too different from the experimental data to be represented on the figures.

Figure 5.12 presents results obtained for the mixture of a 1-1 salt with a 2-1 salt $(NaCl/MgCl_2)$ at a total concentration of 0.5 mol/L. The curvature of the experimental points is due to the variation of the ionic strength. At this concentration of 0.5 mol/L the ideal and limiting law models are also out of the frame of the figure. As the result obtained with $MgCl_2$ is very sensitive to the chosen radii, important differences between the two curves obtained with the cristallographic radii and with this set of radii was observed in the left part of the figure where the $MgCl_2$ con-

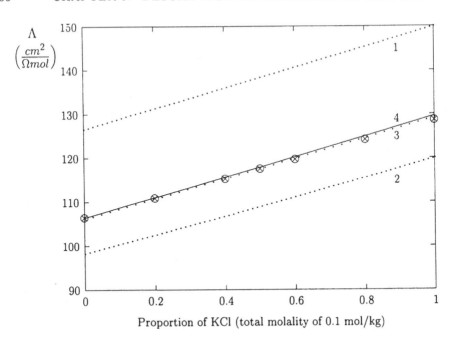

Figure 5.9: Conductance of the NaCl/KCl mixture for a total molality of 0.1 mol/kg.
($\otimes$) : experimental data [33], 1 : Ideal case, 2 : Limiting law (Onsager), 3 : theoretical values predicted by the MSA expression with cristallographic radii, 4 :theoretical values predicted by the MSA expression with the set of radii: $r_{Na^+} = 1.3\text{Å}$, $r_{K^+} = 1.58\text{Å}$ and $r_{Cl^-} = 1.81\text{Å}$.

centration is higher than that of NaCl. As might be expected, the Mg^{2+} radius is very large compared with its crystallographic radius due to its important hydration shell. At lower total concentration, the curves merge again.

Taking the chosen set of radii, very good agreement between theory and experiment was obtained without the need of introducing the concept of ion association for the description of the variation of the conductivity with concentration of an electrolyte solution in the case of 3 different simple ionic species (strong electrolytes). This model provides analytical expressions which are easy to use. However, above the limit of 1 mol/L in total concentration, its validity becomes questionable. A further extension of the theory should involve a modification in the equilibrium model. One possibility would be the use of the HNC model or of other improvements of MSA (softs- MSA, exp- MSA, ...). The problem is then the connection to the low concentration (limiting laws) and the increase in adjustable parameters. Moreover,

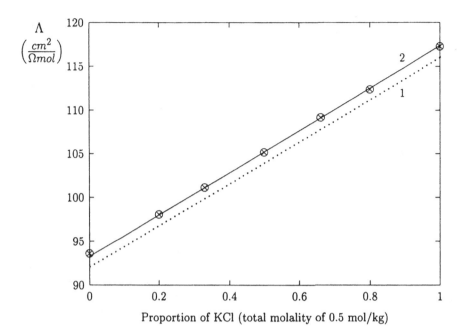

Figure 5.10: Conductance of the NaCl/KCl mixture for a total molality of 0.5 mol/kg. ($\otimes$) : experimental data [40], 1 : theoretical values predicted by the MSA expression with cristallographic radii, 2 : theoretical values predicted by the MSA expression with the set of radii: $r_{Na+} = 1.3\text{Å}$, $r_{K+} = 1.58\text{Å}$ and $r_{Cl-} = 1.81\text{Å}$.

at those concentrations (up to 1M), peculiar physico-chemical phenomena such as ion association could appear.

5.5 Conductance in electrolyte mixtures: Case of micellar systems

An interesting application is the description of the conductivity of charged micellar systems, where, for concentrations above the critical micellar concentration (cmc), at least three species are present in the solution. It is well known that surfactants in water form aggregates above the cmc. In the case of ionic surfactants, below the cmc, the solution consists of surfactant monomer ions and their counterions. Above the cmc, there is an effective loss of ionic charges through ion condensation onto the micellar surface. three types of charged species may be then considered in the

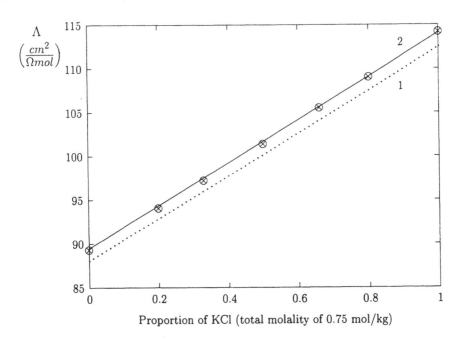

Figure 5.11: Conductance of the NaCl/KCl mixture for a total molality of 0.75 mol/kg. ($\otimes$) : experimental data [40], 1 : theoretical values predicted by the MSA expression with cristallographic radii, 2 : theoretical values predicted by the MSA expression with the set of radii: $r_{Na^+} = 1.3\text{Å}$, $r_{K^+} = 1.58\text{Å}$ and $r_{Cl^-} = 1.81\text{Å}$.

solution: the surfactant monomer, the free counterions not bound to the micellar particles and the micelles. The consequence of the formation of micelles is a sudden change in the conductivity versus surfactant concentration. This change indicates the cmc and the ratio of the linear portion of the curve above and below the cmc is proportional to the degree of counterion condensation. Up to the present time, theories used for the surfactant conductivity were not adequate since they were restricted to the small concentration domain (usually below 0.01 M). This problem was partially solved using better pair distribution functions and the concentration domain could be extended. It is now possible to describe the conductivity of this type of ionic mixtures below and above the cmc.

An important problem in the case of micellar solutions is that the various species have very different sizes and electric charges. The MSA may not be valid under such conditions as concentration is increased. Finally, the introduction of an excluded volume in the relaxation terms instead of a real hard-sphere interaction potential

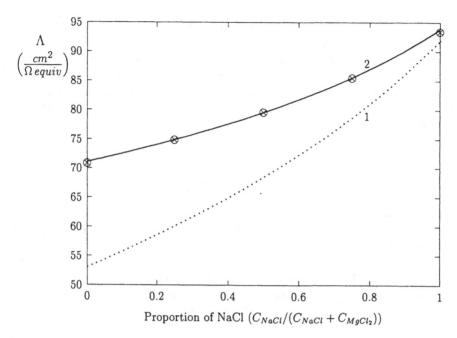

Figure 5.12: Equivalent conductance of the mixture NaCl/MgCl$_2$ at 25°C for a total concentration of 0.5 mol/l (C_{NaCl} + C_{MgCl_2}) versus the NaCl proportion ($C_{NaCl}/(C_{NaCl} + C_{MgCl_2})$). ($\otimes$) : experimental values taken from ref [52]. 1 : theoretical values predicted by the MSA expression with cristallographic radii. 2 : theoretical values predicted by the MSA expression with radii of $r_{Mg^{2+}} = 1.82$Å, $r_{Cl^-} = 1.81$Å and $r_{Na^+} = 1.30$Å.

may be inadequate. Being aware of these limitations, a comparison between the previous model to experimental data available in the literature for various ionic surfactants are presented: three anionic surfactants: sodium dodecylsulfate (SDS) up to a concentration of 0.1 mol/L, sodium octylsulfate up to 0.8 mol/L, sodium octanoate up to 1.5 mol/L as well as results concerning a cationic surfactant: dodecyltrimethylammonium bromide (DTAB) up to 0.1 M. At the highest concentration, the solute volume fraction is 0.1, a value for which the approximations should still be valid.

The first problem is to choose the proper parameters. The data are represented the data as conductance versus the total concentration of monomers as it is done by most authors. The latter quantity is the only one that can be experimentally controlled. All other possibilities depend upon the type of micelle, the micelle aggregation number and the degree of counterion condensation. The conductance is the ratio of the conductivity to the total monomer concentration. Results using MSA are compared with the ideal conductance Λ_{id} and Onsager's result. The ideal conductance is defined as the sum of the conductance of the ions at infinite dilutions multiplied by the concentration of each ionic species and divided by the total monomer concentration c_{mon}^t,

$$\Lambda_{id} = \left(\sum_{i=1}^{3} c_i \lambda_i^o \right) / c_{mon}^t \tag{5.59}$$

Onsager's conductance is that calculated from Onsager's theory [4]. The diffusion coefficient at infinite dilution D_i^o, or its conductance at infinite dilution λ_i^o, for each ionic species and the electrolyte minimum distance of approach σ were used. For the counterions, Na$^+$ and Br$^-$, these two parameters are known. For the monomer surfactant, reasonable radii are chosen. The assumption is that these radii are close to the hydrodynamic radii extracted from the monomeric diffusion coefficient using the approximation of perfect sticking. Then, for an ion j :

$$r_j^{hyd} = k_B T / (6 \pi \eta_o D_j^o) \tag{5.60}$$

where η_o is the viscosity of the pure water. The diffusion coefficients at infinite dilution can either be taken from the literature or considered as adjustable parameters. The hydrodynamic radii are also deduced from the diffusion coefficient through eq.5.60. In effect, below the cmc, besides the minimum distance of approach, the diffusion coefficient at infinite dilution is the only unknown in these expressions. Concerning micelles (above the cmc), in addition to the diffusion coefficient and its minimum distance of approach, the aggregation number (the number of monomer per micelle), a quantity which may vary with concentration, and the apparent charge, which is directly related to the degree of ion condensation, must be known. In order to simplify the model, we assume that all these parameters remain constant as the surfactant concentration varies. Then, the concentration of the various ionic constituents: monomers, micelles, counterions are easily defined.

The surfactants used are 1-1 electrolytes, so that, below the cmc, the monomer concentration is equal to the counterion concentration. Above the cmc, any additional surfactant was considered to be incorporated to the micelles, as required by the pseudo-phase model of micelle formation. Experimentally, however, the monomer concentration c_{mon} is known to decrease somewhat with surfactant concentration above the cmc. In this treatment it is assumed that as micelles are formed, the monomer concentration remains constant whereas a fraction of the counterions will condense onto the micelles.

If n_{agg} is the micelle aggregation number and Z^{app} the apparent charge of the micelle, the concentrations of the various constituents are:

$$c_{mon} = cmc$$

$$c_{mic} = (c^t_{mon} - cmc)/n_{agg}$$

$$c_{counterion} = cmc + \mid Z^{app} \mid c_{mic}$$

where $c_{counterion}$, is the counterion concentration, c_{mic} is the micellar concentration and c^t_{mon} is the total monomer concentration.

The values of all the parameters used are found on figure captions.

Fig. 5.13 shows a comparison between theory and experiment for the cationic surfactant. The experimental results are very reasonably reproduced. Only D^o_{mon} was fitted. All the other parameters were taken from Walrand et al. [53]. They were deduced from quasi-elastic light-diffusion experiments.

Fig. 5.14 presents the same comparison for the anionic surfactant, SDS. The experiment values were taken from the literature [54]. We observe again a very good agreement between theory and experiment. No adjustable parameter was necessary here since the D^o values for monomers and micelles were available from Lindman et al. [55]. These data had been obtained from Fourier Transform Proton NMR experiments (^{1}H FT NMR) and tracer-diffusion.

Fig. 5.15 presents the results obtained for sodium octylsulfate. This surfactant has a high cmc ($\approx 0.135 M$). The agreement between the theory and experiment [57] is satisfactory below the cmc. Above the cmc, experimental and calculated values deviate rapidly. Finally Fig. 5.16 presents the results obtained for sodium octanoate as a function of total monomer concentration. The cmc is very high ($\approx 0.4 M$) and there is also a divergence between experiment and theory above the cmc. There is one fitting parameter: the charge of the micelle.

The effect of changing the values of the different parameters which have to be introduced in the theory is interesting. Small variations have little consequences on the shape of the curves. The largest effect arises from changes in the aggregation number. Experimentally it has been shown [59] that for SDS, the aggregation number varies from approximately 70 at the cmc to 150 at 0.1M. We find, theoretically, that at this concentration the conductance decreases from 18.3 $cm^2\Omega^{-1}mol^{-1}$ for an

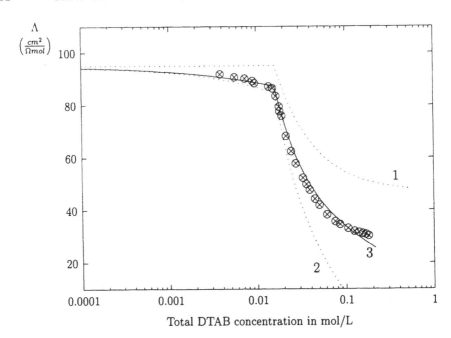

Figure 5.13: Conductance of dodecyltrimethylammonium bromide (DTAB) as a function of total monomer concentration (cmc = 0.016 mol/L). ($\otimes$) : experimental data (personnal experiments). 1 : Ideal case, 2 : Limiting law (Onsager), 3 : Theoretical results (MSA). <u>Parameters :</u> $D^o_{mon} = 4.5 \, 10^{-10} \, m^2 s^{-1}$ (obtained through a fitting procedure), $D^o_{Br^-} = 2.079 \, 10^{-9} \, m^2 s^{-1}$, $D^o_{mic} = 1.15 \, 10^{-10} \, m^2 s^{-1}$ (from reference [53]). Aggregation number of the micelles $n_{agg} = 60$ (from reference [53]). Apparent charge $Z^{app} = +18$ (from reference [53]). $r^{hyd}_{mon} = 5.4$ Å, $r_{Br^-} = 2.17$ Å, $r^{hyd}_{mic} = 21.3$ Å.

aggregation number of 70 to 12 $cm^2 \Omega^{-1} mol^{-1}$ for $n_{agg} = 150$. As all other parameters remaining unchanged, the increase of n_{agg} reflects the decrease of the micelle concentration. Experimentally it is observed that n_{agg} varies less with concentration for anionic micelles than for cationic ones. This may be the consequence of the somewhat better description of the DTAB than of SDS system. This result was predictable. The cmc of DTAB is equal to 0.016 M, whereas that of SDS is 0.0081 M. Our description of the surfactant behavior below the cmc raises no problem, contrary to the micelle/counterion interaction.

Thus the theoretical treatment is certainly better for higher cmc.

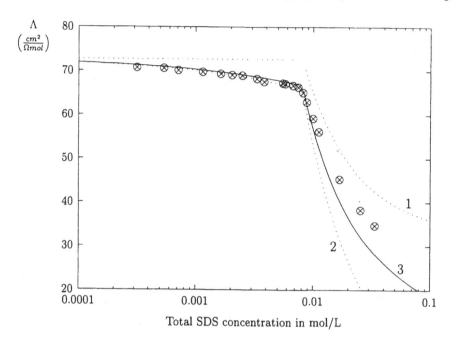

Figure 5.14: Conductance of sodium dodecylsulfate (SDS) as a function of total monomer concentration (cmc = 0.0081 mol/L). ($\otimes$) : experimental data (from reference [54]). 1 : Ideal case, 2 : Limiting law (Onsager), 3 : Theoretical results (MSA). <u>Parameters</u> : $D^o_{mon} = 6.0 \ 10^{-10} \, m^2 s^{-1}$ (from reference [55]), $D^o_{Na^+} = 1.333 \ 10^{-9} \, m^2 s^{-1}$, $D^o_{mic} = 1.15 \ 10^{-10} \, m^2 s^{-1}$ (from reference [55]). Aggregation number of the micelles $n_{agg} = 70$ (from reference [56]). Apparent charge $Z^{app} = -18$ (from reference [56]). $r^{hyd}_{mon} = 4.1$ Å, $r_{Na^+} = 1.07$ Å, $r^{hyd}_{mic} = 21.3$ Å.

Direct application of the theory developed for three simple ions to ionic micellar solutions stresses the following points.

(i.) The description, albeit restricted, is reasonable provided that it is applied to low volume fractions and/or large cmc's and that the interactions between surfactant and counterion are predominant.

(ii.) At the highest surfactant volume fractions, the structure of the micellar system changes (increase in micellar size, transition from spherical to cylindrical symmetry). Micelle/micelle interactions and micelle/counterion interactions become important and the approach departs too much from reality.

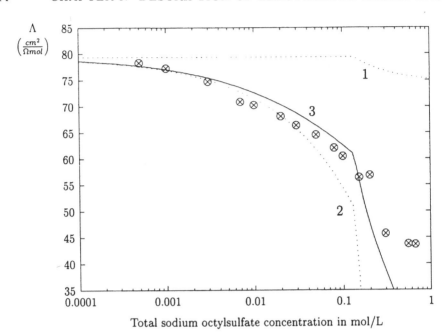

Figure 5.15: Conductance of sodium octylsulfate as a function of total monomer concentration (cmc = 0.135 mol/L). ($\otimes$) : experimental data (from reference [57]). 1 : Ideal case, 2 : Limiting law (Onsager), 3 : Theoretical results (MSA). Parameters : $D_{mon}^o = 7.8 \ 10^{-10} \, m^2 s^{-1}$ (obtained through a fitting procedure), $D_{Na+}^o = 1.333$ $10^{-9} \, m^2 s^{-1}$, $D_{micelle}^o = 2.2 \ 10^{-10} \, m^2 s^{-1}$ (obtained through a fitting procedure) Aggregation number of the micelles $n_{agg} = 24$ (from reference [56]). Apparent charge $Z^{app} = -10$ (from reference [56]). $r_{mon}^{hyd} = 3.1$ Å, $r_{Na+} = 1.07$ Å, $r_{mic}^{hyd} = 11.15$ Å.

This latter point is known and has been addressed before [60, 61]. However, a correct description of the system implies not only improvements by taking into consideration the hard-sphere terms but also the variation of the charge, of the aggregation number and consequently on the radius of the micelle. If such changes can be made, then a finer description of the effect of ionic strength may be expected. As it stands today, this theory describes satisfactorily the concentration zone below and slightly above the cmc. An obvious application of this calculation is the determination of the surfactant diffusion coefficient below the cmc and, at or slightly above the cmc, the evaluation of the micelle diffusion coefficient, the aggregation number or the apparent charge. The latter quantity is generally obtained, using conductance

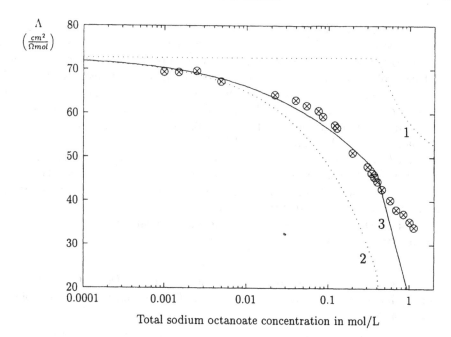

Figure 5.16: Conductance of sodium octanoate as a function of total monomer concentration (cmc = 0.4 mol/L). ($\otimes$) : experimental data (personnal experiments). 1 : Ideal case, 2 : Limiting law (Onsager), 3 : Theoretical results (MSA). Parameters : $D^o_{mon} = 6.0 \ 10^{-10} \, m^2 s^{-1}$ (from reference [58]) $D^o_{Na^+} = 1.333 \ 10^{-9} \, m^2 s^{-1}$, $D^o_{micelle} = 2.0 \ 10^{-10} \, m^2 s^{-1}$ (from reference [58]). Aggregation number of the micelles $n_{agg} = 15$ (from reference [58]). Apparent charge $Z^{app} = -7$ (obtained through a fitting procedure). $r^{hyd}_{mon} = 4$ Å, $r_{Na^+} = 1.07$ Å, $r^{hyd}_{mic} = 12.3$ Å.

versus concentration plots, as recalled above, from the ratio of the slopes below and above the cmc. This evaluation implies an ideal situation (no interaction). Using these same experimental quantities this approach should give an analytical expression for a more realistic calculation of the apparent micellar charge.

5.6 Acoustophoresis of simple salts

In previous transport phenomena, the time dependence was neglected. Concerning acoustophoresis, this approximation is not valid cause the signal originates from inertial effects. Acoustophoresis was originally described by Debye in 1933 [5]. Basically, the phenomenon can be pictured as follows: when an ultrasonic wave is applied to a liquid, the dilation/compression induces the motion of the solvent molecules. In a 1-1 electrolyte solution for instance, the two ions move differently in the solvent's velocity field, because they have different masses and frictional coefficients. This causes local heterogeneities of charges. This effect creates a macroscopic electric field and associated potential differences can be measured in the solution. An illustration is given in the next picture (Fig. 5.17).

Since 1933, acoustophoresis has been extensively studied from the experimental point of view, in the case of electrolyte solutions [62, 63] and colloids [64]. Also, commercial apparatuses have been devised for the purpose of measuring surface electric charges of colloids (ζ-potential) [65].

Since the basic Debye's treatment was based on an ideal solution approximation, workers attempted to take into account the various departures from ideal behaviour [66, 67, 68], namely:

a- For electrolyte solutions, corrections, as applied in the Debye-Onsager theory of conductance [17], were introduced [66]. This approach is questionable for two reasons:

$i-$ the expressions are derived from the Debye-Hückel theory, and thus they are valid only up to 10^{-2}M for 1-1 electrolytes and below for 2-1 salts,

$ii-$ phenomenological electrophoretic and relaxation corrections were introduced in the equations of the acoustophoretic effect with the only justification that, since the observed effect is an electric potential, it involves the same kind of corrections as for electric conductance. These phenomenological corrections do not take into account the specific features of acoustophoresis.

b- Concerning colloidal suspensions, the cell model introduced to describe the effect of ionic interactions is based on the solution of the Poisson-Boltzmann equation around a central colloid particle [69]. Although such a model describes properly counterion-polyion interactions, it is less convincing for polyion-polyion interactions, and does not hold at high polyion molar fractions.

Therefore, it appeared interesting to revisit the theory of acoustophoresis, starting with IVP. Besides, at a practical level, a precise calculation of IVP is useful because IVP can contribute appreciably to the acoustophoresis signal when a mixture of a colloidal suspension with a salt is studied.

Moreover, it is a good example of a time dependent phenomenon. For the same reasons as before, the MSA is used to describe the equilibrium. To understand the physical phenomenon, we present first the Debye's approach of the acoustophoresis.

Ultrasonic Vibration Potential

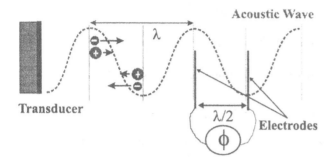

Electrokinetic Sonic Amplitude

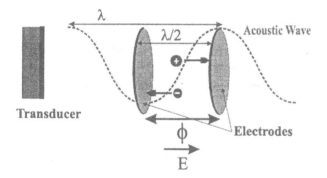

Figure 5.17: Principle of acoustophoresis. Concerning the UVP (Ultrasonic Vibration Potential - top), an ultrasonic wave applied on a liquid (transducer) induces solvent motion. As the two charged species have a different masses and frictional coefficients, its move differently. The charge heterogeneousness which appeared in this way generate a macroscopic and thus measurable electric field (electrodes). Concerning the ESA (Electro Sonic Amplitude - bottom), an alternative electric field is applied (electrodes). Eachs ion species moves in opposite direction. This motion induces a detectable ultrasonic wave (transducer).

5.6.1   Debye's treatment

In 1933 Debye [5] postulated that the application of a sinusoidal acoustic wave to an electrolyte solution gives rise to a local potential Φ

$$\Phi = \Phi_o \exp\left[i\left(\omega t - px - \Delta\right)\right] \tag{5.61}$$

where x is the direction of propagation, ω is the pulsation of the wave, p is the wave number and Δ is the phase shift.

The basic ingredients of Debye treatment are the following: Newton's law,

$$F_j^{el} + F_j^{fr} = m_j \frac{dv_j}{dt} \tag{5.62}$$

is written **for** each ion j, where m_j is the apparent mass of the ion j, with

$$F_j^{el} = e_j E \tag{5.63}$$

the electric force, with e_j the charge of the ion j and **E** the local electric field.

Moreover

$$F_j^{fr} = -\zeta_j(v_j - v_s) \tag{5.64}$$

is the friction force, with ζ_j the friction coefficient of the particle j, v_s the velocity of the solvent and v_j the velocity of the particle j.

Combining eq.(5.62) with the hydrodynamic continuity equation for each ion:

$$\frac{\partial c_j}{\partial t} + \frac{\partial(c_j v_j)}{\partial x} = 0 \tag{5.65}$$

where c_j is the concentration of species j, and with the Poisson equation (in cgs units)

$$\frac{\partial E}{\partial x} = \frac{4\pi}{\varepsilon} \sum_j c_j e_j \tag{5.66}$$

where ε is the relative dielectric constant of the solvent and the propagation direction is along x. From these equations,

$$E = \frac{4\pi v_s}{\varepsilon k_B T} \left[\sum_{j=1}^{2} \bar{c}_j e_j m_j \left(D_j^0 - \frac{k_B T D_j^0}{g_s^2 m_j}\right)\right] \Big/ \left[1 - i\frac{4\pi}{\varepsilon k_B T \omega} \sum_{j=1}^{2} D_j^0 \bar{c}_j e_j^2\right] \tag{5.67}$$

with $D_j^0 = k_B T / \zeta_j$ the diffusion coefficient of the species j at infinite dilution given by the Nernst-Einstein relation.

And using the fact that

$$E = -\frac{\partial \Phi}{\partial x} = ip\,\Phi \tag{5.68}$$

the potential amplitude follows:

$$\Phi_o = g_s v_{so} \left| \frac{4\pi}{\omega \varepsilon} \sum_{j=1}^{n} \frac{\bar{c}_j e_j m_j}{\zeta_j} \right| \Big/ \left[1 + \left(\frac{4\pi}{\omega \varepsilon} \sum_{j=1}^{n} \frac{\bar{c}_j e_j^2}{\zeta_j} \right)^2 \right]^{1/2} \tag{5.69}$$

where n is the number of ionic species, g_s is the sound velocity, v_{so} the solvent velocity amplitude which depends on the power of the acoustic ultrasonic wave and $\bar{c}_j$ is the equilibrium concentration of species j.
For a 1-1 salt, the potential Φ_o becomes:

$$\Phi_o = c \frac{\Psi \, \Delta\tau}{\omega} \Big/ \left[1 + \left(\frac{1}{\omega \tau_D} \right)^2 \right]^{1/2} \tag{5.70}$$

where c is the equilibrium concentration and τ_D is the Debye relaxation time defines as

$$\tau_D = \left[\frac{4\pi \, c\,e^2}{\varepsilon} \left(\frac{1}{\zeta_+} + \frac{1}{\zeta_-} \right) \right]^{-1} \tag{5.71}$$

and

$$\Delta\tau = |\tau_+ - \tau_-| \tag{5.72}$$

where τ_+ and τ_- are the inertial times respectively of the cation and of the anion

$$\tau_j = \frac{m_j}{\zeta_j} \tag{5.73}$$

with $j = +, -$ and

$$\Psi = \frac{4\pi \, g_s v_{s0}\, e}{\varepsilon} \tag{5.74}$$

with e the charge of the proton.
The potential given by eq.(5.69) is concentration dependent: at low concentration, it increases with the concentration and above some critical concentration (above $10^{-3}M$) it reaches a plateau value.
Eq.(5.70) shows that there are two limiting cases:

i. when $\omega \tau_D \gg 1$ (low c), $\Phi_o \approx c \Delta \tau \psi / \omega$

ii. when $\omega \tau_D \ll 1$ (high c), $\Phi_o \approx c \tau_D \Delta \tau \Psi$.

Consequently, at large c (low $\omega \tau_D$) Φ_o is independent of c, which yields the plateau value.

For instance, ir the case of $NaCl$, we have that $\tau_D \approx 5.5 \, 10^{-11}/c$ (in s with c in mol/L) $\tau_{Na^+} = 1.23 \, 10^{-11}$ s, $\tau_{Cl^-} = 2.91 \, 10^{-11}$ s, at $\omega = 1.26 \, 10^6$ s^{-1}, $\lambda = 0.7$ cm and $g_s = 1.4 \, 10^5$ cm/s.

So, Debye's treatment leads to a plateau for the potential Φ_o, at high c. However, it is observed experimentally that some salts do not exhibit any plateau, but rather a continuously decreasing pattern. As the frequency is typically on the order of 100 kHz for a wave lenght of 1 cm and a pressure amplitude around 0.1 Atm in water, absorption effect or the ionic association could not explain this decreasing. The consideration of non ideal terms gives an explanation of this behaviour.

5.6.2   Calculation of non ideal terms

In addition to relaxation and electrophoretic forces (fig. 5.1), which are corrections to the main driving forces, the diffusion must be taken into account. Indeed, acoustic wave induces concentration gradients and then, diffusion effect appears. This effect will be taken into account after the calculation of ralaxation and hydrodynamic corrections. MSA is again used to describe the equilibrium state.

Relaxation

The main driving force to be included in the calculation of the relaxation effect is the electric force. Following Onsager's formalism [2], and assuming that the non-equilibrium time-dependent two particle correlation function is of the form

$$\hat{h}_{21}(r, t) = h^0_{21}(r) + h'_{21}(r) \exp(i \omega t) \tag{5.75}$$

where h^0_{21} stands for the equilibrium two particle correlation function, it is found that $h'_{21}(r)$ obeys

$$\Delta h'_{21} - \kappa^2_q h'_{21} = \mathbf{A}_1 \nabla h^0_{21} \tag{5.76}$$

with

$$\kappa^2_q = \frac{4\pi}{\varepsilon} \frac{n_1 e^2_1 a_1 + n_2 e^2_2 a_2}{D^0_1 + D^0_2} + i \frac{\omega}{D^0_1 + D^0_2} \tag{5.77}$$

$$\mathbf{A}_1 = \frac{1}{D_1^0 + D_2^0} \left[a_1 (e_1 \mathbf{E}(\mathbf{r}_1) + \zeta_1 \mathbf{v}_s(\mathbf{r}_1)) - a_2 (e_2 \mathbf{E}(\mathbf{r}_2) + \zeta_2 \mathbf{v}_s(\mathbf{r}_2)) \right] \tag{5.78}$$

where n_i is the mean number density, D_i^0 is the self-diffusion coefficient, e_i is the charge, $\mathbf{r}_1$ and $\mathbf{r}_2$ are the positions of ions 1 and 2 and

$$a_i = (\zeta_i + i\, m_i \omega)^{-1} \tag{5.79}$$

for each species i.

Lastly, h_{21}^0 is replaced by its MSA expression [70, 26, 27], which leads after some algebra, and expanding the results in a first-order power series of $m_i \omega / \zeta_i$ (typically of the order of 10^{-8}), to the relaxation correction force

$$\delta \mathbf{k}_i^{rel} = e_i\, \alpha_k\, \mathbf{E} + i\, e_i\, \omega\, \beta_k\, \mathbf{v}_s \tag{5.80}$$

with

$$\alpha_k = \frac{\mathrm{Re}(\kappa_q^2)}{3} \left(i_0(\kappa_q \sigma) - \frac{\epsilon k_B T}{e_i e_j} \kappa_q \sigma^2 . i_1(\kappa_q \sigma) \right) \frac{\kappa_q B_{ij} \exp(-\kappa_q \sigma)}{\kappa_q^2 + 2\Gamma \kappa_q + 2\Gamma^2 (1 - \exp(-\kappa_q \sigma))} \tag{5.81}$$

$$\begin{aligned}
\beta_k ={}& -\frac{4\pi \sum_{i=1}^{2} \bar{c}_i e_i m_i D_i^0}{3\epsilon\, k_B T \sum_{i=1}^{2} D_i^0} \left(i_0(\kappa_q \sigma) - \frac{\epsilon k_B T}{e_i e_j} \kappa_q \sigma^2 . i_1(\kappa_q \sigma) \right) \\
&\times \frac{\kappa_q B_{ij} \exp(-\kappa_q \sigma)}{\kappa_q^2 + 2\Gamma \kappa_q + 2\Gamma^2 (1 - \exp(-\kappa_q \sigma))}
\end{aligned} \tag{5.82}$$

with

$$B_{ij} \simeq \frac{e_i e_j}{\epsilon k_B T (1 + \Gamma \sigma)^2} \tag{5.83}$$

where $\mathrm{Re}(z)$ is the real part of the complex z, Γ is the MSA parameter, κ the inverse Debye screening length, σ the mean closest-approach distance

$$i_0(\kappa_q \sigma) = \frac{\sinh(\kappa_q \sigma)}{\kappa_q \sigma} \tag{5.84}$$

$$i_1(\kappa_q \sigma) = \frac{\cosh(\kappa_q \sigma)}{\kappa_q \sigma} - \frac{\sinh(\kappa_q \sigma)}{\kappa_q^2 \sigma^2} \tag{5.85}$$

Hydrodynamic correction

The velocity increment δv_j^{hyd} induced on a particle j by the motion of the surrounding particles i can be found by the use of the Oseen tensor [31]. The result reads

$$\delta v_j^{hyd} = \sum_{i=1}^{2} n_i \int_{\mathbf{r}} h_{ji}^0(\mathbf{r})\, \mathcal{T}(\mathbf{r})\, \zeta_i(v_i - v_s)\, d\mathbf{r} \qquad (5.86)$$

where $\mathcal{T}$ is the Oseen tensor given by:

$$\mathcal{T}(\mathbf{r}) = \frac{1}{8\pi\eta r}\left(\mathcal{U} + \frac{\mathbf{r}\otimes\mathbf{r}}{r^2} \right) \qquad (5.87)$$

with $\mathcal{U}$ the unit tensor and η the viscosity of the solvent.
Finally:

$$\delta v_j^{hyd} = e_j\, \alpha_v\, \mathbf{E} + i\, e_j\, \omega\, \beta_v\, \mathbf{v}_s \qquad (5.88)$$

with

$$\alpha_v = -\frac{\kappa}{6\pi\eta\,(1 + \Gamma\sigma)^2} \qquad (5.89)$$

$$\beta_v = \frac{2}{3\,\eta\,\epsilon\, k_B T\,(1 + \Gamma\sigma)^2} \sum_{i=1}^{2} \bar{c}_i e_i m_i/\kappa \qquad (5.90)$$

and the electrophoretic force is given by:

$$\delta k_j^{hyd} = \zeta_j\, \delta v_j^{hyd} \qquad (5.91)$$

Result

The diffusion force is taken as

$$F_j^{dif} = -k_B T\frac{\partial \ln(c_j)}{\partial x}\, \mathbf{x} \qquad (5.92)$$

which, setting $c_i = \bar{c}_i + \delta c_i$ with $\delta c_i \ll \bar{c}_i$, becomes

$$F_j^{dif} = i\, \frac{k_B T}{\bar{c}_i}\, p\, \delta c_i\, \mathbf{x} \qquad (5.93)$$

where $\mathbf{x}$ is the unit vector in the propagation direction x.
Now, the various corrections are included in eq.(5.62), namely:

$$m_i \frac{dv_i}{dt} = \mathbf{F}_j^{el} + \mathbf{F}_j^{fr} + \mathbf{F}_i^{dif} + \delta \mathbf{k}_i^{hyd} + \delta \mathbf{k}_i^{rel}$$

$$i\omega m_i v_i = e_i \mathbf{E} - \zeta_i (v_i - v_s - \delta v_i^{hyd}) - \frac{k_B T}{\bar{c}_i} \frac{\partial c_i}{\partial x} \mathbf{x} + \delta \mathbf{k}_i^{rel} \qquad (5.94)$$

and the same procedure as followed by Debye [5] is used, which yields after some calculations

$$E = \frac{4\pi v_s}{\epsilon k_B T} \left[\sum_{i=1}^{2} \bar{c}_i e_i m_i \left(\omega_i - \frac{k_B T D_i^0}{g_s^2 m_i} \right) \right] \Bigg/ \left[1 - i \frac{4\pi}{\epsilon k_B T \omega} \sum_{i=1}^{2} \bar{c}_i e_i^2 \omega_i \right] \qquad (5.95)$$

with

$$\omega_i = D_i^0 (1 + \alpha_k) \left[1 - \frac{k_B T \kappa}{6\pi\eta D_i^0 (1 + \Gamma\sigma)^2} \right] \qquad (5.96)$$

Where α_k is given by eq.(5.81). Eq.(5.96) is similar to eq.(5.67) found by Debye, with ω_i in lieu of D_i^0, except for the diffusion correction.
Using eq.(5.68), it follows that the potential amplitude is given by

$$\Phi_o = \frac{4\pi g_s v_{so}}{\omega \epsilon k_B T} \left(\frac{\left[\sum_{i=1}^{2} \bar{c}_i e_i m_i \left(\mathsf{Re}(\omega_i) - \frac{k_B T D_i^0}{g_s^2 m_i} \right) \right]^2 + \left[\sum_{i=1}^{2} \bar{c}_i e_i m_i \mathsf{Im}(\omega_i) \right]^2}{\left[1 + \frac{4\pi}{\epsilon k_B T \omega} \sum_{i=1}^{2} \bar{c}_i e_i^2 \mathsf{Im}(\omega_i) \right]^2 + \left[\frac{4\pi}{\epsilon k_B T \omega} \sum_{i=1}^{2} \bar{c}_i e_i^2 \mathsf{Re}(\omega_i) \right]^2} \right)^{1/2}$$

$$(5.97)$$

where $\mathsf{Im}(z)$ is the imaginary part of the complex z.

Eq.(5.97) contains three unknowns: the closest-approach distance σ, the mass of the solvated anion, and that of the solvated cation.
We can see in eq.(5.97) that in the denominator of Φ_o, the main contribution is the specific conductance, as in the Debye's treatment. This quantity is known with a very good precision. In order to have a model consistent with this quantity, the closest-approach distance σ is calculated by fitting the conductance data of a given salt by the expression for the equivalent conductivity

$$\Lambda_{eq} = \sum_{i=1}^{2} \lambda_i^0 (1 + \alpha_k') \left[1 - \frac{k_B T \kappa}{6\pi\eta D_i^0 (1 + \Gamma\sigma)^2} \right] \qquad (5.98)$$

where $\alpha_k' = \mathsf{Re}(\alpha_k)$ and λ_i^0 is the equivalent conductance at infinite dilution for the species i, and it is given by

$$\lambda_i^0 = \frac{D_i^0 |z_i| F^2}{RT} \qquad (5.99)$$

where F is the Faraday constant, z_i the charge number of ion i and R is the gas constant.

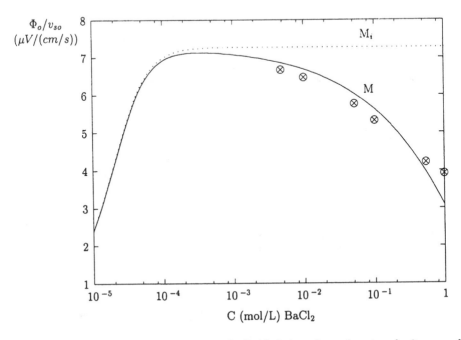

Figure 5.18: Acoustophoretic potential divided by the solvent velocity ampli-tude against BaCl$_2$ salt concentration. $\otimes$: Millner's experimental results [63]: $f = 170\,\text{kHz}$ and $T = 22\,^\circ C$. M: fitted curve with a mass for Ba^{2+} of 304 g/mol (hydration number of 9.3) and M$_{Cl^-}$ = 35.5 g/mol. Mi: corresponding ideal curve.

Eq.(5.98) is derived from the expression reported in ref. [13] which contained individual cut-off distances and second-order terms. It is a simplification of the latter in the sense that it is limited to the first-order term and it is written as a function of a mean cut-off distance. It is this simplified expression which was used in the subchapter " Conductance of two simple ionic species ".
Concerning the masses of the ions, it can be noticed that, after the original work of Debye, some authors [67, 68] proposed to correct the masses from a buoyancy effect. In the equation of motion of an ion a term was added to allow for a pressure gradient on an ion. However a question arises about this correction. Although it would be suitable for particles of macroscopic size, like colloids, its applicability to

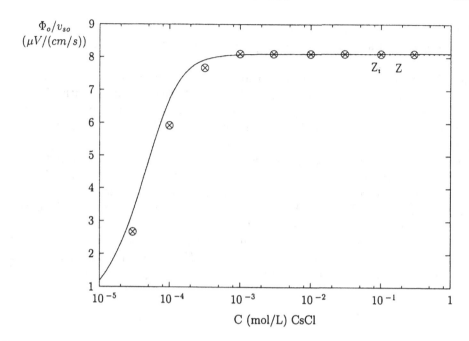

Figure 5.19: Acoustophoretic potential divided by the solvent velocity amplitude against CsCl salt concentration. $\otimes$: Zana's experimental results [71]: $f = 220$ kHz and $T = 25\,°C$. Z: fitted curve with a mass for Cs^+ of 146 g/mol (hydration number of 0.6) and $M_{Cl^-} = 35.5$ g/mol. Zi: corresponding ideal curve.

simple ions is questionable because the latter have a size comparable to that of the water molecules. Some authors [71] related the apparent mass of the ions W_i to the volume of the ions and that of the bound solvent. They expressed therefore their results in terms of ionic partial molar volumes. Here, results are described in term of solvation numbers, which may be compared to other results obtained from other techniques. Moreover, it may be expected that the rigorous correction should be smaller than that used previously (buoyancy correction). In the absence of a reliable expression not to introduce any correction originating from the pressure gradient effects was preferred. This work describes the motion effects in terms of an effective mass $M_i^{app} = M_i^0 + M_i^s$ where M_i^0 is the mass of the bare ion and M_i^s is the mass of solvent bound to the ion.

We present just two results from ref. [12]. The two salts are BaCl$_2$ and CsCl where the chloride ion is assumed un-hydrated [72, 73, 74]. The only fitting parameter is then the hydration number of the cation.

The plots of the scaled signal Φ_o/v_{so} are given as a function of the concentration for each salt.

The plots relative to $BaCl_2$, show an appreciable decrease in this concentration range. The data are fitted with eq.(5.97). In fact, the agreement with Millner's experimental results [63] is excellent. Like conclusions can be drawn in the case of other 2-1 salts ($SrCl_2$ and $CaCl_2$): good fit of Millner's experimental results has been obtained [12].

Results concerning CsCl exhibit a clear plateau. Zana's data [71] include measurements at very low CsCl concentration, down to 10^{-5} mol/L. Surprisingly good agreements of theoretical result with the experimental data is found, even at low CsCl concentration. The fact that no decrease of the signal occurs at high concentration originates from the closeness of the cation's and anion's diffusion coefficients. This particularity makes the hydrodynamic and relaxation corrections almost equal for each ion, as can be seen in eq.(5.95), the total effect negligible. The same effect is found for RbCl salt wherereas other studied 1-1 salts exhibit decreases [12, 63, 71].

Bibliography

[1] L. Onsager *Phys. Z.* 27 (1926) 388.; *ibid.* 28 (1927) 277.

[2] L. Onsager and R.M. Fuoss *J. Phys. Chem.* 36 (1932) 2689.

[3] L. Onsager *Ann. N. Y. Acad. Sci.* 46 (1945) 263.

[4] L. Onsager and S.K. Kim *J. Phys. Chem.* 61 (1957) 215.

[5] P. Debye *J. Chem. Phys.* 1 (1933) 13.

[6] P. Debye and E. Hükel *Phys. Z.* 24 (1923) 185.

[7] P. Turq *Chem. Phys. Lett.* 15 (1972) 579.

[8] C. Micheletti and P. Turq *J. Chem. Soc. Faraday. Trans II* 73 (1976) 743.

[9] W. Ebeling and J. Rose *J. Solution Chem.* 10 (1981) 599.

[10] W. Ebeling and M. Grigo *J. Solution Chem.* 11 (1982) 151.

[11] O. Bernard, W. Kunz, P. Turq and L. Blum *J. Phys. Chem.* 96 (1992) 398.

[12] S. Durand-Vidal, J.-P. Simonin, P. Turq, and O. Bernard

[13] O. Bernard, W. Kunz, P. Turq, and L. Blum *J. Phys. Chem.* 96 (1992) 3833.

[14] P. Turq, L. Blum, O. Bernard and W. Kunz *J. Phys. Chem.* 99 (1995) 822.

[15] A. Chhih, P. Turq, O. Bernard, J.M.G. Barthel and L. Blum *Ber. Bunsen. Phys. Chem.* 12 (1994) 1516.

[16] R.A. Robinson and R.H. Stokes in *Electrolyte Solutions*, Butterworth, London, 1959.

[17] P. Debye and H. Falkenhagen *Physikalische Zeitschrift* 13 (1928) 401; P. Debye and H. Falkenhagen *Zeitschrift fur Physikalische Chemie A* 137 (1928) 399.

[18] O. Bernard, P. Turq and L. Blum *J. Phys. Chem.* 95 (1991) 9508.

[19] P. Turq, J. Barthel and M. Chemla *Transport, Relaxation, and Kinetic Processes in Electrolyte Solutions*, Lecture Notes in Chemistry, Springer, Berlin Heidelberg, 1992.

[20] W. Ebeling, R. Feistel, G. Kelbg and R. Sändig *J. Non-Equilibrium Thermodynamics* 3 (1978) 11.

[21] P. Résibois *Electrolyte Theory*, Harper and Row Publishers, New York, Evanston and London, 1968.

[22] J. C. Rasaiah and H. L. Friedman *J. Phys. Chem.* 72 (1968) 3352.

[23] Yu. V. Kalyuzhnyi, V. Vlachy, M. F. Holovko and G. Stell *J. Chem. Phys.* 102 (1995) 5770.

[24] J.K. Percus and G. Yevick *Phys. Rev.* 110 (1958) 1.

[25] E. Waisman and J. L. Lebowitz *J. Chem. Phys.* 56 (1971) 3086.

[26] L. Blum *J. Mol. Phys.* 30 (1975) 1529.

[27] L. Blum and J.S. Hoye *J. Phys. Chem.* 81 (1977) 1311.

[28] C. Sánchez-Castro and L. Blum *J. Phys. Chem.* 93 (1989) 7478.

[29] T. Cartailler, P. Turq, L. Blum and N. Condamine *J. Phys. Chem.* 96 (1992) 6766.

[30] W. Kunz, P. Turq, M.-C. Bellissent-Funel and P. Calmettes *J. Chem. Phys.* 95 (1991) 6902.

[31] H.L. Friedman *Physica* 30 (1966) 537.

[32] J. Quint and A. Viallard *J. Chim. Phys.* 69 (1972) 1095.

[33] J. Quint *Thèse de Doctorat d'Etat*, Université de Clermont-Ferrand, France, 1976.

[34] S. Durand Vidal *Thèse de Doctorat de Physique de l'Université P.&M. Curie*, Paris 6, France, 1995.

[35] J. Anderson and R. Paterson *J. Chem. Soc. Faraday I* 71 (1975) 1335.

[36] *International Critical Tables of Numerical data, Physics, Chemistry and Technology.* published for the National Research Concil by the McGRAW-HILL Book Compagny Inc., 1927.

[37] V.M.M. Lobo *Electrolyte solutions, Data on Thermodynamic and Transport Properties*, edited by Coimbra Editora, Portugal Vol. I, 1984.

[38] T. Shedlovsky and A.S. Brown *J. Am. Soc.* 56 (1934) 1066.

[39] J.F. Chambers, J.M. Stokes and R.H. Stokes *J. Phys. Chem.* 60 (1956) 985.

[40] D. Urban, A. Geret, H.-J. Kiggen and H. Schönert *Z. für Phys. Chem.* 137 (1983) 66.

[41] T. Shedlowski *J. Amer. Chem. Soc.* 54 (1932) 1411.

[42] J.F. Chamber, J. M. Stokes, R. H. Stokes *J. Phys. Chem.* 60 (1956) 985.

[43] P.C. Carman *J. Phys. Chem.* 73 (1969) 1095.

[44] D.G. Miller *J. Phys. Chem.* 70 (1966) 2639.

[45] M. Postler *Coll. Czech. Chem. Commun.* 35 (1970) 535.

[46] A.N. Campbell and L. Ross *Can. J. Chem.* 34 (1956) 566.

[47] C.G. Swain and D.F. Evans *J. Amer. Chem. Soc.* 88 (1966) 383.

[48] J.C. Justice *J. Chim. Phys.* 65 (1968) 353.

[49] T. Shedlovsky and A.S. Brown *J. Amer. Chem. Soc.* 56 (1934) 1066.

[50] T.L. Broadwater and D.F. Evans *J. Sol. Chem.* 3 (1974) 757.

[51] S. Phang, R. H. Stokes *J. Sol. Chem.* 9 (1980) 497.

[52] H. Bianchi, H.R. Corti and R. Fernandez-Prini *J. Sol. Chem.* 18 (1989) 485.

[53] S. Walrand, L. Belloni and M. Drifford *J. Physique* 47 (1986) 1565.

[54] B.D. Folckhart and H. Graham *J. Coll. Sci.* 8 (1953) 428.

[55] B. Lindman, M.-C. Puyal, N. Kamenka, R. Rymden and P. Stilbs *J. Phys. Chem.* 88 (1984) 5048.

[56] D.F. Evans and B.W. Ninham *J. Phys. Chem.* 87 (1983) 5025.

[57] F.D. Haffner, G.A. Piccionne and C. Rosenblum *J. Phys. Chem.* 46 (1942) 662.

[58] P. Turq, M. Drifford, M. Hayoun, A. Perera and J. Tabony *J. Physique lett.* (1983) L-471.

[59] J.B. Hayter and J. Penfold *J. Chem. Soc. Faraday Trans I* 77 (1981) 1851.

[60] E. Y. Sheu, C.-F. Wu, S.-H. Chen, and L. Blum *Physical Review A* 32 (1985) 3807.

[61] J.P. Hansen and J.B. Hayter *Molecular Physics* 46 (1982) 651.

[62] R. Zana and E. Yeager *Mod. Aspects. Electrochem.* 14 (1982) 1.

[63] R. Millner *Z. Elektrochem.* 65 (1961) 639.; R. Millner and H.-D. Müller *Ann. Phys. (Leipzig)* 17 (1966) 160.

[64] B. J. Marlow, D. Fairhurst and H.P. Pendse *Langmuir* 4 (1988) 611.

[65] J. S. Heyman *U.S. Patent* (1991) No. 7-628062

[66] S. Oka *Proc. Phys.-Math. soc. Jap.* 13 (1933) 413.

[67] J. J. Hermans *Phil. Mag.* 25(7) (1936) 426.

[68] J. Bugosh, E. Yeager and F. Hovorka *J. Chem. Phys.* 15 (1947) 592.

[69] S. Alexender, P. M. Chaikin, P. Grant, G. J. Morales, P. Pincus and D. Hone *J. Chem. Phys.* 80 (1984) 5776.

[70] E. Waisman and J.L. Lebowitz *J. Chem. Phys.* 52 (1970) 4307.

[71] R. Zana and E. Yeager *J. Phys. Chem.* 70 (1966) 954.

[72] J.F. Hinton and E.S. Amis *Chemical Reviews* 71(6) (1971) 627.

[73] G.J. Herdman and G.W. Neilson *J. Mol. Liq.* 46 (1990) 165.

[74] A.P. Copestake, G.W. Neilson and J.E. Enderby *J. Phys. C: Solid State Phys.* 18 (1985) 4211.

Chapter 6

Polyelectrolytes

6.1   Introduction

Polyelectrolytes are highly charged nanoscopic objects or macromolecules. Their electric charge density appears as more or less continuous, when it is seen from distances to the macromolecule equal to several times to the intercharge distance, giving them the polyelectrolytic character. Obviously, their properties will be extremely different according to their geometry. Massive spherical objects will behave like colloids, whereas linear flexible objects will keep some of the macromolecular polymeric character.

In the present short chapter only linear polyelectrolytes will be considered, since the globular objects will be analogous, as regards their interfacial electrolytic properties, to charged colloids, which are examined in other chapters.

As a first approximation the charge effects of coulomb interactions are taken into consideration in their short range character which appears in the so called condensation phenomena [1]. Experimental examples are reported in [7].

The polyelectrolyte solution is composed of: polymers, counterions and salt. Examples: polyacids (DNA), polybases (PSSNa, PSSTMA), proteins above or below their isoelectric point. These examples give an idea of the huge biological importance of polyelectrolytes.

Polyelectrolytes are by no way a mere superposition of electrolytes and polymers properties. New and rather unexpected behaviours are observed:

- Whereas polymers exhibit only excluded volume effects, the long ranged coulomb ionteractions, which are present in polyelectrolytes give rise to new critical exponents.

- The main difference with electrolytes is that one kind of ions, the counterions are stuck together along a chain, the collective contribution of the charged monomers causes a strong field in the vicinity of the chain, even at very low dilution.

6.2   counterion condensation

In this very basic presentation Manning-Onsager's argument will be followed.

Consider an infinitely thin rod , whose linear charge density is $\lambda = -e/A$. There may be some salt added and the valence of the counterions is $Z > 0$. Let call r_n the distance of the nth counterion from the rod and look at a configuration where the 1st counterion is at $r_1 < r_0 << \kappa^{-1}$ whereas any other counterion is farther than κ^{-1}.

For $r < r_0$ the screening is negligible so that the potential energy of the 1st counterion is merely

$$E_p(r_1) = \frac{2Ze^2}{4\pi\varepsilon_0\varepsilon_r A} \ln(r_1)$$

For $j \geq 2$ the screened potential is

$$E_p(r_1) \simeq K_0(\kappa r_j)$$

where K_0 is the modified Bessel function and κ is the Debye Hckel parameter

$$\kappa^2 = 4\pi Q (\sum_a n_a Z_a^2)$$

and

$$Q = \frac{\beta e^2}{4\pi\varepsilon_0\varepsilon_r}$$

the sum runs over each ionic species a.

The partition function of the system is

$$Z_N = \int dr_1... \int dr_N \exp[-\beta \sum_{j=1}^{N} E_p(r_j)]$$

$$> \int r_1 dr_1 \exp[-\beta E_p(r_1)][\int_{r_0}^{\infty} r dr \exp(-\beta E_p(r))]^{N-1}$$

$$= f(r_0) \int_0^{r_0} r_1^{1-2Z\xi} dr_1$$

with $\xi = \frac{Q}{A}$ is the ratio of the Bjerrum length by the charge separation.

If $2Z\xi > 2$, the partition function Z_N is diverging, so that for $\xi > 1/Z$ the system presents a phase transition.

For water at $25C$, one has $Q = 714pm$

Manning assumes that condensation of counterions occurs to prevent the divergence of this partition function. The net result will be to reduce the apparent charge density on the polyion, until the value of ξ after condensation will be equal to the critical value $1/Z$. In the case of DNA, two consecutive pairs of phosphates are separated by $A = 170pm$, then $\xi = 4.2$, so that any ion, regardless to its charge will condense on DNA, until the effective value of ξ will be reduced to $1/Z$.

This very simple argument leads to prediction in good agreement with most experimental data and can be generalized for more complicated situations by integrating the Poisson Boltzmann equation with the proper symmetry.

The main criticisms on this simple model concern

- the behaviour in very dilute solution where entropy considerations would imply some kind of mass action law giving some decondensation.

- the fact that real polyelectrolytes have a finite diameter. The previous treatment does not apply anymore in this case.

Anyway, the success of Manning-Onsager's argument makes this model very popular and shows the predominant role of Coulomb interactions for polyelectrolytes.

The introduction of ionic interactions for polyelectrolyte solutions, at least at the Debye-Hückel level, is another problem, which can be considered either for equilibrium or for dynamic properties.

Recent important progresses have been made in this direction, for the theory of static and equilibrium properties of polyelectrolytes by Barbosa and Levin. The interested reader will find all required details which are out the scope of this chapter in the original papers. [2] [3] [4] [5] [6].

It should be noticed that those approaches do not apply to either transport or dynamic applications, whereas MSA and related tools have been applied to counterion and polyion dynamics [7] [8] [9] [10] [11].

Bibliography

[1] J.M. Victor *Basic Theory of Polyelectrolytes in The Physics and Chemistry of Aqueous Ionic Solutions*, M.C Bellissent-Funel and G.W. Neilson editors, NATO ASI Series, Reidel, Dordrecht, 1987.

[2] M.N. Tamashiro, Y. Levin and M.C. Barbosa *Physica A.* 258 (1998) 341.

[3] Y. Levin, M.C. Barbosa and M.N. Tamashiro *Europhys. Let.* 41 (1998) 123.

[4] P.S. Kuhn, Y. Levin and M.C. Barbosa *Macromolecules* 31 (1998) 8347.

[5] A. Diehl, M.C. Barbosa and Y. Levin *Phys. Rev. E* 54 (1996) 6516.

[6] Y. Levin and M.C. Barbosa *J. Phys. II* 7 (1997) 37.

[7] P. Turq, J. Barthel and M. Chemla *Transport, Relaxation and Kinetic Processes in Electrolytes* Lecture Notes in Chemistry, 57, Springer Verlag, Berlin, 1992.

[8] O. Bernard, P. Turq and L. Blum *J. Phys. Chem.* 95 (1991) 10576.

[9] S. Durand Vidal, J.P. Simonin, P. Turq and O. Bernard *Progr. Colloid. Polym. Sci.* 98 (1995) 184.

[10] P. Turq, O. Bernard, J.P. Simonin, S. Durand-Vidal, J. Barthel and L. Blum *Ber. Bunsenges. Phys. Chem.* 100 (1996) 738.

[11] A. Chhih, P. Turq, O. Bernard, J. Barthel and L. Blum *Ber. Bunsenges. Phys. Chem.* 98 (1994) 1516.

List of Figures

List of Tables

Index